Nitrogen and Climate Change

Nitrogen and Climate Change

An Explosive Story

Dave Reay
Professor of Carbon Management and Assistant Principal,
Global Environment & Society, University of Edinburgh, UK

palgrave
macmillan

First published 2015 by
PALGRAVE MACMILLAN

Palgrave Macmillan in the UK is an imprint of Macmillan Publishers Limited, registered in England, company number 785998, of Houndmills, Basingstoke, Hampshire RG21 6XS.

Palgrave Macmillan in the US is a division of St Martin's Press LLC, 175 Fifth Avenue, New York, NY 10010.

Palgrave Macmillan is the global academic imprint of the above companies and has companies and representatives throughout the world.

Palgrave® and Macmillan® are registered trademarks in the United States, the United Kingdom, Europe and other countries.

ISBN: 978-1-137-28694-9   hardback
ISBN: 978-1-137-28695-6   paperback

This book is printed on paper suitable for recycling and made from fully managed and sustained forest sources. Logging, pulping and manufacturing processes are expected to conform to the environmental regulations of the country of origin.

A catalogue record for this book is available from the British Library.

Library of Congress Cataloging-in-Publication Data

Reay, Dave, 1972–
     Nitrogen and climate change : an explosive story / Dave Reay, Professor of Carbon Management and Assistant Principal, Global Environment & Society, University of Edinburgh, UK.
        pages   cm
     Includes bibliographical references.
     ISBN 978-1-137-28695-6 (paperback) — ISBN 978-1-137-28694-9 (hardback)
     1. Nitrogen—Environmental aspects.   2. Climate change mitigation.   3. Nitrogen cycle.   I. Title.
     TD885.5.N5R425 2015
     551.51'12—dc23   2015002666

In memory of Phyllis, Frank and Bernard Hutchinson

Contents

List of Table, Figures and Boxes

Table

Figures

Boxes

Preface – Wonderstuff

The faces say it all. Tell people you work on climate change and they'll have an opinion. They'll comment on their memories of childhood weather, the price of oil and the hand-wringing convenience of air travel. Tell them you work on nitrogen and at best you'll get a blank look, more often it's a poorly disguised yawn. The very word 'Nitrogen' sounds flat, an echo of radio gardening programmes and the more mundane corners of the periodic table.

Had history turned out differently, and its discoverer not literally lost his head, this forgettable geek of the elemental year book would instead go by the zippy name of Azote – the life-taker. Instead, we have nitrogen. Common as muck, most often lacking any smell, taste or colour, and the science writing equivalent of a sports book on golf club grips, there would seem little point in telling its story. But, just beneath this dull veneer lies the stunning truth about a world-changing substance. Scratch the surface of the global challenge that is climate change, peer further into the perfect storm of population growth, food shortages and water pollution, and it is the layered and interconnected threads of nitrogen that shine through. Running through life, death and decomposition, integral to our genes, the food we eat, the air we breathe and the climate impacts we face – it is a wonderstuff.

**

Our high-school chemistry class was not sent wild with excitement the first time we heard about nitrogen. It was spring, and through the dark months of winter the stained and malodorous Mr Davies had dragged us on through the hinterlands of the periodic table. For hour after hour we scribbled down the key facts about hydrogen, helium and the rest and, if we were lucky, got to try out the peculiarities of these different elements in the lab. We had made hydrogen go 'pop' in a test tube, had been roundly disciplined for breathing in helium from balloons to make our voices squeak, and had burnt some coal to heat some water. For nitrogen though, there was to be no practical. What was there to do with this odourless, colourless and unreactive gas but just write down that it was as common as muck and didn't do a whole lot. We had oxygen coming up the following week and the promise of setting things on fire, so for now it was just a case of fidgeting in the clammy plastic chairs and hoping someone would fart to break the monotony.

The key facts on nitrogen were dutifully regurgitated onto exam papers at the end of the year and that, as far as we were concerned, was pretty much that. My own plan was to be a marine biologist – to discover new species, walk the beaches of the world and dive to unexplored ocean depths. Never again would I have to recall where nitrogen came in the periodic table and all the things it didn't do.

I thought nitrogen didn't matter. That it was irrelevant to the intensifying global challenges of water, food, energy and climate security, that the important things in my life – family, friends and a sustainable future – had nothing to do with nitrogen. On every count, I was wrong.

A life in nitrogen

One of my earliest recollections is of a chill Guy Fawkes's night at my grandparents' house in Billingham in North East England. Taking turns with my brother to select the smooth, brightly coloured tubes from the box of fireworks that evening was a joy that burned into my memory as surely as the explosions of white stars left their afterglow behind my eyelids. That it was nitrogen I had to thank for the giggle-inducing display (Box P.1) would have meant nothing to me at the time, yet as we crowded together for warmth our family stood at the British epicentre of humankind's manipulation of this potent substance.

My grandparents' house and those around were built for workers in the huge nitrogen factory that loomed over the town. When the wind blew from its towering stacks the washing was brought in and the windows shut as the rich and pungent fumes spread down the streets. The town lived and breathed nitrogen – even its football team 'the Synthonia' was named after the process of manufacturing the stuff.

My grandfather Frank was an obsessive gardener and made full use of the free bags of fertiliser that came his way from the factory. His large garden was a riot of productivity, the frequent drenching of nitrogen from the air combining with the handfuls he cast across the soil to power incredible plant growth and ensure a bumper harvest each year. His one disappointment, and a casualty of the gardening success everywhere else, was his pond. Whatever he did, no matter how many times it was cleaned out, within a week or two it had always become an opaque soup of green algae – these microscopic

---

Box P.1

Rocket fuel

Following their 9th century discovery of gunpowder – made from a form of reactive nitrogen called nitrate, mixed with charcoal and sulphur – the Chinese soon realised that if the explosive force of this black powder could be directed as a high speed jet in one direction then it could make whatever it was contained in fly very fast the other way. This fiery demonstration of Newton's third law – that every action has an equal and opposite reaction – has been streaking the world's skies ever since. Wonderful for fireworks it may have been, but as a weapon it was initially more effective in creating shock and awe than as a means of delivering directed destruction. Achieving any great degree of accuracy required improved materials and launching systems and it wasn't until the Second World War that the use of rockets on the battlefield really took hold.

By then the smoky, inefficient, gunpowder-propelled rockets had long been replaced by more powerful and more accurate versions that could be fired rapidly and in enormous numbers. The real breakthrough for rocket technology came with designs that could combine the fuel and the source of its ignition – called the oxidiser, and usually in a form of reactive nitrogen called ammonium perchlorate – together in exactly the right proportions while it was in flight. With the greater control and power this combination offered, the range of rockets began to stretch far beyond the normal limits of the battlefield. As the tide was turning against the Germans in the Second World War they pressed ahead with the development of a rocket with a range sufficient to deliver a high explosive warhead to the homes and factories of southern England. These so-called V2 rockets were the first man-made objects to enter space, their flight taking them to heights of over 100 miles above Earth before beginning the speeding descent back down towards London or Paris.

---

plants feasting on the helpings of airborne nitrogen even more greedily than his cabbages.

Both my parents had grown up in this nitrogen-producing world capital. War loomed large over the family for decades after its end, the large painting of my Uncle Ronnie – killed in fighting near Arnhem – and the defused bomb that lay somewhere deep beneath Frank's weighty rows of cabbages, reminders of the explosive role nitrogen played in this chapter of world and family history.

Nitrogen with friends

My childhood dream to dive to unexplored depths began to take shape at university. As trainee scuba divers, we were schooled in the effects of pressure on the human body as well as the usual rules about hand signals and safety checks.

During the downward plunge the pressure of the air breathed by the diver increases rapidly and so too does the amount of nitrogen gas that passes into the blood stream. As the level ramps up, it blocks more and more of the messages passed from one nerve cell to another. On the brain the effects are not unlike those of alcohol, and the deeper a diver goes the more noticeable the impact, each 10 metres deeper purportedly being the equivalent of knocking back another Martini. Suffering from such narcosis – or getting 'narced' as it is more commonly described – can be great fun, but when you are deep beneath the sea being drunk on nitrogen is risky in the extreme. Badly affected divers have been known to take out their air regulator and offer it to passing fish, or sit giggling on the seabed as their tank ran out.

For the university sub-aqua club, the pleasures and dangers of becoming narced were a continual source of fascination. The idea of a hangover-free binge had its attractions, but we novices were wisely limited to shallow dives in the murk of the local docks, with any inebriation reserved for the warming shots of whisky in the post-dive pub stop. Our chance to sample the delights of narcosis eventually came in the safer and entirely water-free environment of a recompression chamber. Dressed only in shorts and t-shirts, we crowded into its spartan confines and took our seats for the descent. The grins were initially strained as the pressure was slowly ratcheted up, but soon the surge of nitrogen in our bloodstreams made itself felt. Euphoria swept through the chamber and, for reasons that remain unclear, our voices all took on the form of high-pitched Mancunians. At the equivalent of 50 metres underwater and in the full-blown rapture of the deep we urged the stout operator outside to push us even deeper. Clearly inured to exhortations from drunken Mancunians he brought us back up in long, ear-popping stages and as the tide of nitrogen in our bodies flowed back out in our breath the euphoria slipped away. Our chamber-borne trip to the deep was a jolly evening that did little to curb our desire to dive to great depths. But for the chamber's intended users, its cramped confines may mean the difference between life and death (Box P.2).

Box P.2

Nitrogen 'bends'

The flow of nitrogen into and back out of our bodies generally goes unnoticed – any new molecules that pass into the blood stream being exchanged with old ones that have completed their quiet sojourn in the body. But when the pressure of the air we breathe changes radically, as it does for a descending or ascending scuba diver, this quiet shuffling in and out of molecules can quickly become a disorderly and life-threatening rush.

The 'bends' have long been a menace to unwary divers. More properly called decompression sickness, it occurs when the nitrogen-loaded body of the diver rises too fast towards the surface. If the exchange of gas from the blood back into the lungs is too slow then the gas will begin to come out of solution and form bubbles instead. The first signs that these bubbles are forming in the bloodstream are itchy skin and aching joints. Left untreated a 'bend' rapidly intensifies, the excruciating pain causing victims to double over in agony – the characteristic symptom that gives it its name. Staged stops in the ascent from a deep dive are designed to give the gas enough time to flow out via the lungs and so avoid the bloodstream bubbles. If these stops are missed then the only option is to recompress the body in the hope that the bubbles will dissolve.

Nitrogen at work

The mid-1990s found me as a doctoral student aboard a research ship in the Southern Ocean. Our quest was to discover how climate change might affect these vast and violent waters, and almost immediately another glittering strand of nitrogen's key role in life and death began to shine. Not all nitrogen, I soon became aware, was inert. The cold waters that swirled around the ship were teeming with life dependent on getting enough of the stuff. As I simulated one hundred years' of planetary warming in racks of heated water jars, the rampant thirst of all plants and animals for nitrogen, the bursts of growth when it was plentiful and the life or death struggles when it wasn't were played out in miniature before my eyes. It was a revelation – this stuff was actually important. Mr Davies would have been proud.

The more I learned about it, the more I saw just how powerful it was. From the algae-smothered streams down in South Georgia to the terrorist bomb plots uncovered back home in the UK, the negative impacts of nitrogen were increasingly obvious. But then so too were its many benefits. When my wife was due to give birth it was nitrous oxide that was provided to ease the pain, when my father was flattened by an attack of angina it was a spray of nitroglycerine into his mouth that had him up and back digging in the garden within minutes. The downsides of excess nitrogen may have driven my pond-loving granddad to despair, but in fields around the world it was helping to put food in the mouths of billions. This was a substance with myriad forms that had shaped human civilization for millennia and whose fingerprints were apparent on almost every facet of life of Earth.

Alongside these last four decades of personal nitrogen discovery has come the emergence of global climate change as one of the greatest threats our world faces in this century and beyond. Scientific understanding of nitrogen's myriad interactions with climate change is still incomplete, yet the power it can exert for good or ill in a warming world is all too evident. Its story is of the peculiar and the mundane, of water turning red and people turning blue, one of climate friend and pollution foe, of meaty feasts and looming famine. This is nitrogen and climate change.

Acknowledgements

This book has been a long time in the writing. Countless weekend mornings have gone into its creation, mornings where the lure of sunshine in the garden and water fights with the children has often condemned progress to a few damp paragraphs. To Sarah, my wonderful wife, and my daughters, Maddy and Molly, thank you for putting up with my lunchtime grumps about how things like 'nitrogen fixation' could possibly be made to sound interesting.

A wide-ranging examination of nitrogen's interactions with climate change – as attempted here – inevitably drew heavily on the research of colleagues with whom I have worked over the last 20 years. In particular, thank you to Keith Smith, Karen Dobbie, Paul Crutzen, Eric Davidson, Mark Sutton, Stefan Reis, Tony Edwards, Pete Smith, Phil Ineson and Dave Nedwell. You have opened my eyes to the wonders of nitrogen and its potential to help us in the battle against dangerous climate change.

Thank you also to Olivia Eadie, Stuart Sayer, Sarah Ivory and all the graduate 'Carbon Masters' at the University of Edinburgh who have taught me so much more than I ever taught them, and whose great work on climate change around the world inspires me every day.

The tangle of nitrogen's pathways through air, water and land would be impossible to describe in words alone. The various cycles pictured in this book were drawn by me with the help of images obtained from the fantastic 'thenounproject.com'. Some of the images used in these diagrams are made available under a Creative Commons attribution thanks to these designers:

Outfall pipe, dead bird and crop duster designed by Luis Prado
Fish designed by Mallory Hawes
Tree designed by Nicolas Ramallo
Fire designed by John Caserta
Plants designed by Hero Arts and Jerry Wang
Plane designed by Sven Gabriel
Clouds and chemistry designed by Scott Lewis and Zoe Austin
Microbes designed by Maurizio Fusillo and Drue McCurdy

Finally, thank you to my parents and grandparents, for instilling in me a thirst for discovery that has never faded, for those fireworks and giant cabbages, for the holiday songs and the wartime stories. This is for you.

Introduction

Climate change is one of the greatest challenges to humankind in the 21st century, and the writing on it has already swallowed up a good-sized forest of trees in books ranging from climate science to policy, from engineering to economics and from self-help manuals to disaster fiction. Nitrogen – a rather less attention-grabbing subject – inevitably has a far smaller literary canon.

Things are changing though. In climate science writing, nitrogen is getting a nod more frequently – the pages devoted to nitrogen in the door-stopping Intergovernmental Panel on Climate Change (IPCC) assessment report are testament to its role in past, present and future climate. In the small shelf of 'nitrogen books' too, the global web of interactions between nitrogen and the climate is invariably flagged up.

The aim of this book is to overtly bring these civilisation-defining subjects together in one place, to explore the ways in which nitrogen underpins climate change, and to expose the synergies and antagonisms that on one hand could feed billions and on the other could drive catastrophic global change for our species.

This book first delves into the role of nitrogen in human history and the efforts to wrestle control of nitrogen supplies around the world. The initial chapters aim to show just how fundamental nitrogen is to human civilisation and, as such, how resilience to climate change can be either shored-up or undermined by it. Subsequent chapters look at nitrogen's more direct role in global climate. The book takes an in-depth look at nitrous oxide – nitrogen's most obvious face of the greenhouse effect – before examining the interactions of nitrogen and climate change in the air, in freshwaters, in our oceans and on land. Finally, it explores the forces that are set to make nitrogen an even greater driver of global change and the ways in which we might make better use of this dull-named but wonderful substance in the fight to avoid dangerous climate change.

A quick guide to 'Nitrogen and Climate Change'

Nitrogen's many interactions and feedbacks with climate change can be broken down into a few key areas – atmosphere, water and land – and, within these, on how nitrogen and climate combine to affect human and natural systems around the world. As already mentioned, the most obvious links between nitrogen and climate change are via changes in the greenhouse effect (Figure 1.1).

A direct impact of 'nitrogen deposition' (the drizzle of reactive nitrogen contained in dust, snow, rain and aerosols) on climate change is through resulting emissions of the powerful greenhouse gas nitrous oxide from land and oceans where nitrogen is deposited. This enhancement of the greenhouse effect can be somewhat offset (reduced warming) by the deposited nitrogen leading to more plant growth and carbon dioxide uptake, and by its tendency to reduce production of methane (another powerful greenhouse gas).

The addition of nitrogen via fertilisers leads to a similar set of impacts on the global greenhouse effect, though here the nitrogen tends to be added in much greater amounts over more specific areas (i.e. agricultural fields) than the more scattergun mode of atmospheric deposition. Because of the large inputs of nitrogen to soils, the production and then emission of nitrous oxide tends to be the dominant impact on the greenhouse effect here.

Emissions of other forms of nitrogen – such as nitrogen oxides (called NOx) from fossil fuel burning and ammonia (NH_3) from agriculture – can themselves have an indirect effect on the global energy balance. This can be through their involvement in the production of ozone (another greenhouse gas, one that can damage plant growth), the lifetime of methane in the atmosphere or through the formation of aerosols that alter the amount of sunlight that is reflected by our atmosphere.

Nitrogen and climate change in the atmosphere

Nitrogen is a major driver of climate change via the greenhouse effect, but climate change itself can interact with nitrogen in our atmosphere to exacerbate or buffer impacts on society and ecosystems. In our atmosphere, air quality is a prime example of the threat such interactions pose. For instance, air temperatures are projected to increase in many areas. Both higher temperatures and more emission of nitrogen oxides tend to increase

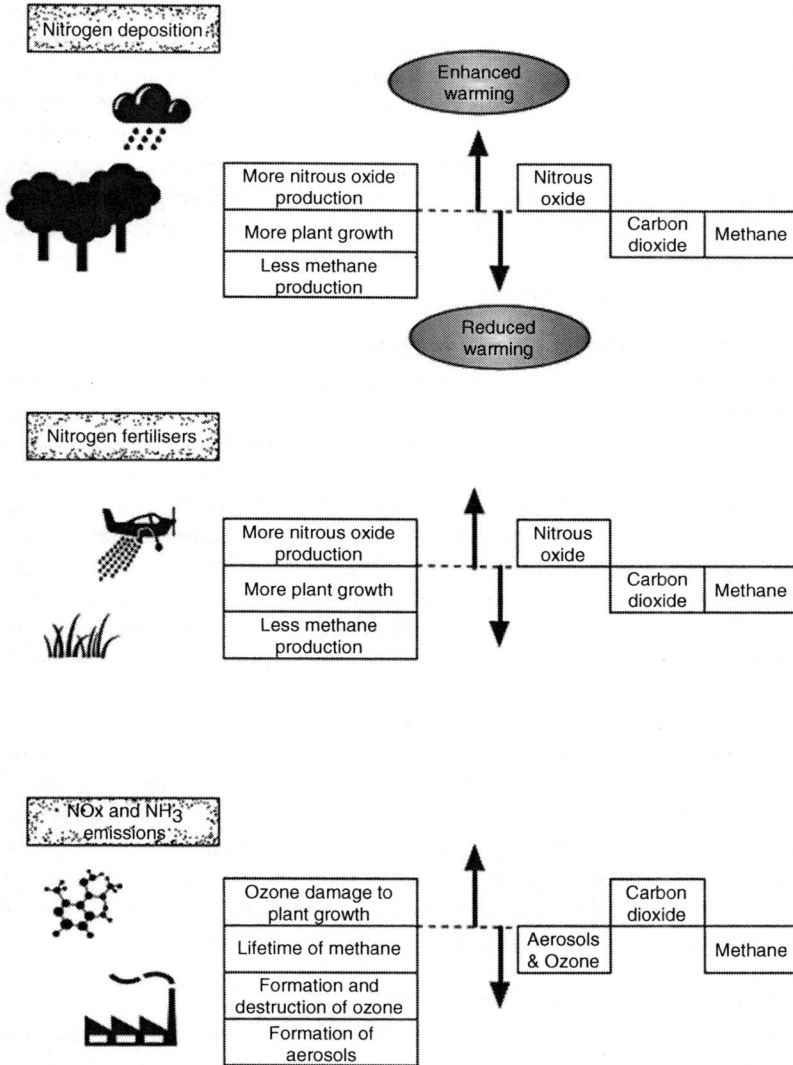

Figure I.1   Key interactions of nitrogen with the global greenhouse effect

For each of the three nitrogen pathways shown – 'deposition', 'fertilisers' and 'NOx and NH$_3$' – the main effects on greenhouse gases are shown in terms of whether they enhance warming (upward arrow) or reduce it (downward arrow).

Source: Dave Reay

the production of low-level ozone – a serious human health risk. So, in combination, nitrogen and climate change can push ozone concentrations up to much more dangerous levels than either could do alone (Figure I.2). Similarly, more nitrogen deposition and carbon dioxide may increase plant growth in many areas, but climate change is also projected to increase the frequency of wildfires. Where these effects coincide, the impact of wildfires in terms of carbon emissions and air pollution may therefore be much greater. Changes in the amounts of pollen and volatile organic compounds (VOCs) emitted into our atmosphere from vegetation can also have serious consequences for air quality.

Nitrogen and climate change in water

For aquatic systems, it is water quality that exemplifies nitrogen and climate change interactions. In both freshwater and marine environments around the world, increasing inputs of nitrogen are causing faster growth of plants and algae (eutrophication). Eutrophication can reduce biodiversity, damage fisheries and promote the growth of toxic algal species. On top of this, climate change is projected to lead to heavier rainfall and increased snow and ice melt in some areas, which means that some rivers, lakes and coastal waters will receive even greater inputs of nitrogen due to run-off – enhancing eutrophication even further. For some stream and river systems, the boost in aquatic plant growth combined with intense rainfall and meltwater flows may block drainage channels and exacerbate flood risk (Figure I.3).

In areas where climate change leads to reduced rainfall, concentrations of dissolved nitrogen in the form of ammonia or nitrate may rise to levels that break rules for drinking water quality or threaten fish health. Likewise, increasing temperatures may reduce mixing of water and exacerbate problems of low oxygen supply (hypoxia) in lakes and oceans. Warmer, more acidic waters are also expected to change the types and distribution of aquatic algae in many areas and so alter nitrogen fixation rates.

Nitrogen and climate change on land

On the land too, changes in temperature and rainfall due to climate change may interact with nitrogen in a multitude of ways. For instance, higher surface temperatures may lead to higher rates of ammonia loss from agriculture (due to volatilisation). Changes in warming and precipitation are also expected to allow some plant species to colonise new areas – putting at risk those species already present. This invasive species effect can then be enhanced by the effects of additional nitrogen inputs and eutrophication that favour the invading plants (e.g. grasses) over the existing ones (e.g. mosses) (Figure I.4).

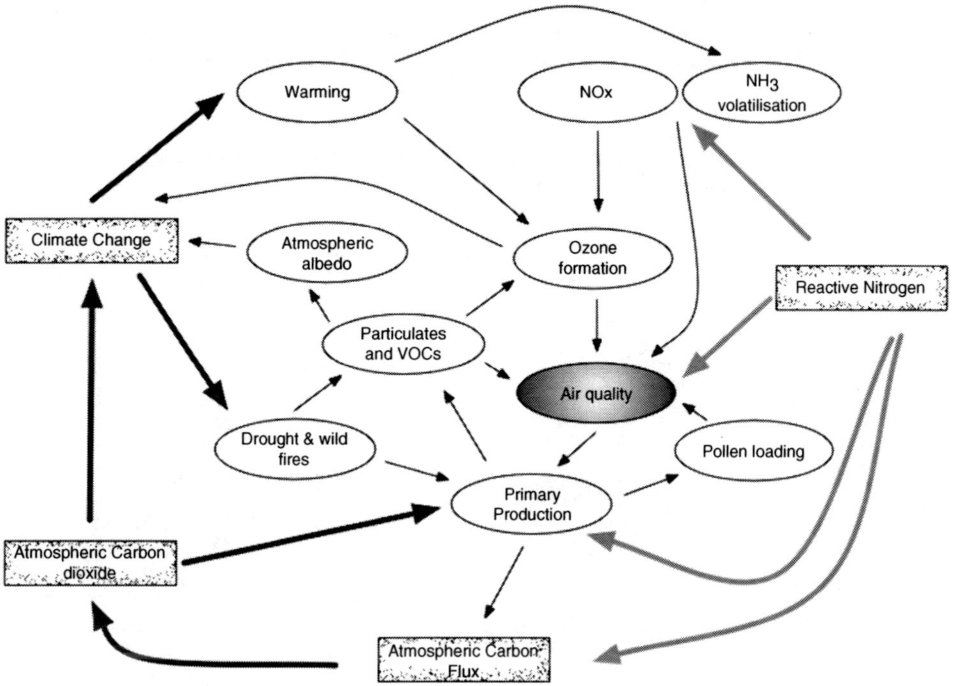

Figure I.2 The web of nitrogen and climate change interactions in our atmosphere

The major impacts of climate change and increasing carbon dioxide concentrations (large black arrows) and the major impacts of nitrogen (large grey arrows) come together directly in the atmosphere through effects such as ozone formation (from higher temperatures and more NOx emissions).

Source: Dave Reay

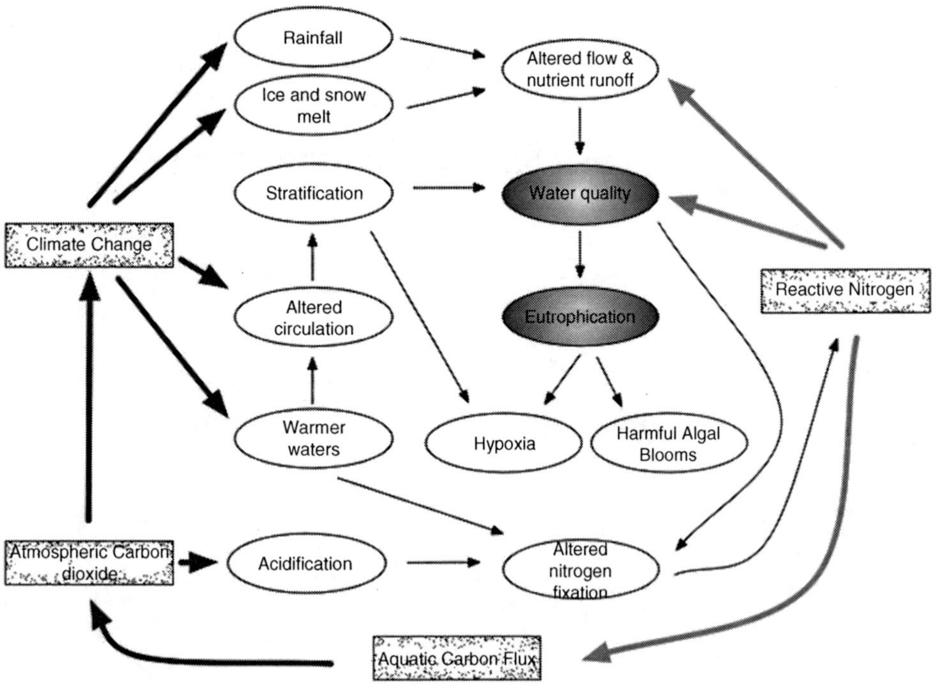

Figure I.3   Key interactions of nitrogen and climate change in aquatic environments

The major impacts of climate change and increasing carbon dioxide concentrations (large black arrows) and the major impacts of nitrogen (large grey arrows) come together directly in the aquatic environment through effects such as poor water quality and eutrophication (from more nitrogen inputs and increased run-off).

Source: Dave Reay

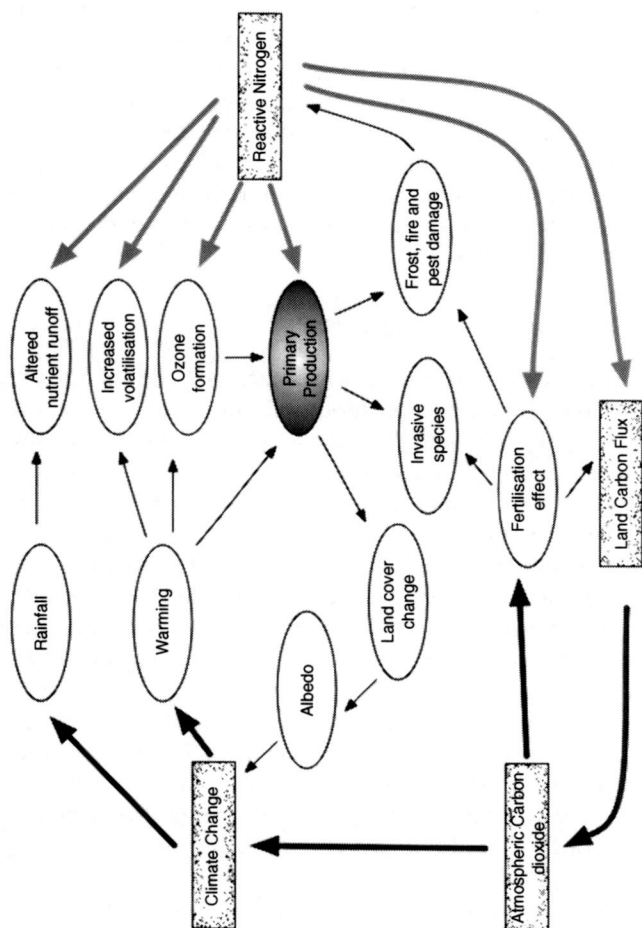

Figure I.4 Key interactions of nitrogen and climate change on land

The major impacts of climate change and increasing carbon dioxide concentrations (large black arrows) and the major impacts of nitrogen (large grey arrows) come together directly in the terrestrial environment through interactions such as altered vegetation growth (from more nitrogen inputs and increased carbon dioxide). In some cases the extra growth induced by more nitrogen and carbon dioxide actually makes the vegetation more suscepti-ble to damage by fire, pests and frosts.

Source: Dave Reay

An important nitrogen-climate interaction in terms of limiting global warming is via the carbon dioxide 'fertilisation effect'. As concentrations in the atmosphere rise, plant growth (primary production) in some areas will benefit and more carbon will be taken up. This fertilisation effect is often limited by nitrogen supply, so more nitrogen could allow global vegetation to take up much more carbon and better buffer carbon dioxide emission from human activities. However, this synergistic effect may be undermined by other climate change impacts such as drought, increase in pest attacks, damage from frost and disease, and frequency of wildfires. Finally, changing primary production and vegetation types due to climate change and nitrogen may alter the reflectance (albedo) of land cover. A global increase in albedo (more reflective land cover) would help to limit climate warming, while a transition to darker types of land cover would reinforce it.

Past and current nitrogen management

As the negative impacts of poor nitrogen management and its interactions with climate change have become more and more evident in recent decades, a swathe of policies aimed at addressing it has emerged. At the global level, for instance, the United Nations Framework Convention on Climate Change (UNFCCC) requires nations to report their emissions of nitrous oxide. Where a nation also has climate change mitigation targets, nitrous oxide is included in the so-called 'basket of gases' by which these targets can be met. Likewise, nitrogen trifluoride (NF_3) – a very powerful greenhouse gas used in the manufacture of electronics – has recently been added to the list of greenhouse gases that nations must report and control.

Across Europe, losses of nitrogen to aquatic systems have been targeted through the Common Agricultural Policy (CAP) and specific legislation such as the European Commission (EC) Nitrates Directive. Nitrogen pollution of the air from fossil fuel burning has also been partially addressed through strict limits on emissions from power plants, such as via the EC Large Combustion Plants Directive.

These and other national and regional policies around the world are helping to reduce reactive nitrogen problems to a certain extent. However, the real challenge in addressing nitrogen in a changing climate is that of dealing with so many different forms of nitrogen, in so many different places, and with so many different effects.

1
A Brief History of Nitrogen

It has been 100 years since a German scientist named Fritz Haber came up with a large-scale way to convert more of the sea of nitrogen gas around us into a usable form[1]. Before then, how much food we could produce from the fields was largely down to how well we recycled manure and made use of the nitrogen-fixing magic produced by plants like peas and beans. Haber's invention has allowed us to green the world's increasingly exhausted fields and put food on the table of billions. A staggering two out of every five people alive today are thought to owe their continued existence to his process[2], yet millions still go hungry, and producing enough food for the burgeoning human population of the 21st century will test how well we manage the limit for the use of this precious substance. To date, our record is not a good one.

We have been criminally wasteful in super-charging the fields of the world. The rich doses of nitrogen intended to grow more crops have instead served to contaminate drinking water and power huge blooms of harmful algae in our lakes and seas[3]. In the air, the deliberate enrichment of the land has ramped up levels of the powerful greenhouse gas nitrous oxide[4], while the noxious emissions that are belched from countless power stations, factories and vehicles around the world threaten ever more people with respiratory diseases[5]. For humankind, nitrogen is truly two-faced.

Haber himself may be viewed either as the saviour of billions or as a nationalistic monster whose inventions prolonged the bloodiest of world wars and whose legacy included chemical warfare and Hitler's gas chambers[6]. To tell the true story of nitrogen then is to gaze on both faces of this Janus element – the brutal and the beneficent. It is an integral part of my life, your life and every life on Earth. From food and anaesthesia to space exploration and cryopreservation, from shellfish poisoning and blue baby syndrome, to war, famine and terrorism, its reach is global and yet so often overlooked. Ultimately, whether Haber's gift to the world ends up being a blessing or a curse for future generations will depend on just how well we learn from the past, and that means looking back a very long way indeed.

Little green men

Each molecule of the trillions of tonnes of nitrogen gas that makes up most of the air we breathe has two atoms. These atoms share a triple bond that is one of the strongest found in nature and, for the first billion or so years of our planet's existence, only nature's most powerful atmospheric sledgehammer could break it. Travelling at speeds of up to 10,000 miles per hour and heating the air around it to an air-splitting 25,000°C, the raw energy of lightning is able to tear apart the powerfully bound pairs of nitrogen atoms and combine them with oxygen to form reactive nitrogen oxides[7]. Much of this then ends up dissolved in rainfall, sprinkling usable nitrogen back over the surface of the Earth[8] (Figure 1.1).

For the very first life-forms drifting in the sparse oceans of almost four billion years ago, the nitrogen supplied by lightning was a crucial handhold on existence. Their numbers grew and the grip of life on Earth became more tenacious, but as they used up more and more of the nitrogen supplied by lightning the handhold it provided began to crumble. As demand increased, the supply faltered. The Earth's atmosphere was changing, with concentrations of carbon dioxide in the atmosphere dropping as increasing amounts were locked away in freshly exposed surface layers, and life on Earth placed its first flagella-prints on the climate. By using artificial lightning – super-heated plasma produced with a laser – we now know that in this changing atmosphere, lightning strikes would have produced less and less reactive nitrogen[9]. The earliest life on Earth, it seems, was at risk of tripping over itself in the dash for continued existence.

Competing for a diminishing supply of usable nitrogen and faced with an apparently insurmountable limit to expansion, the niche for organisms able to use nitrogen directly from the atmosphere became a yawning chasm. Evolution, that greatest filler of gaps, did not disappoint. In relatively short order the pocket battleships of life on Earth – the cyanobacteria – had arrived. Cyanobacteria (also known as blue-green algae) may be tiny, but the impact these early photosynthesis-endowed microbes have had on our planet is hard to overstate. These microorganisms were able to convert the inert nitrogen gas into a usable form, to

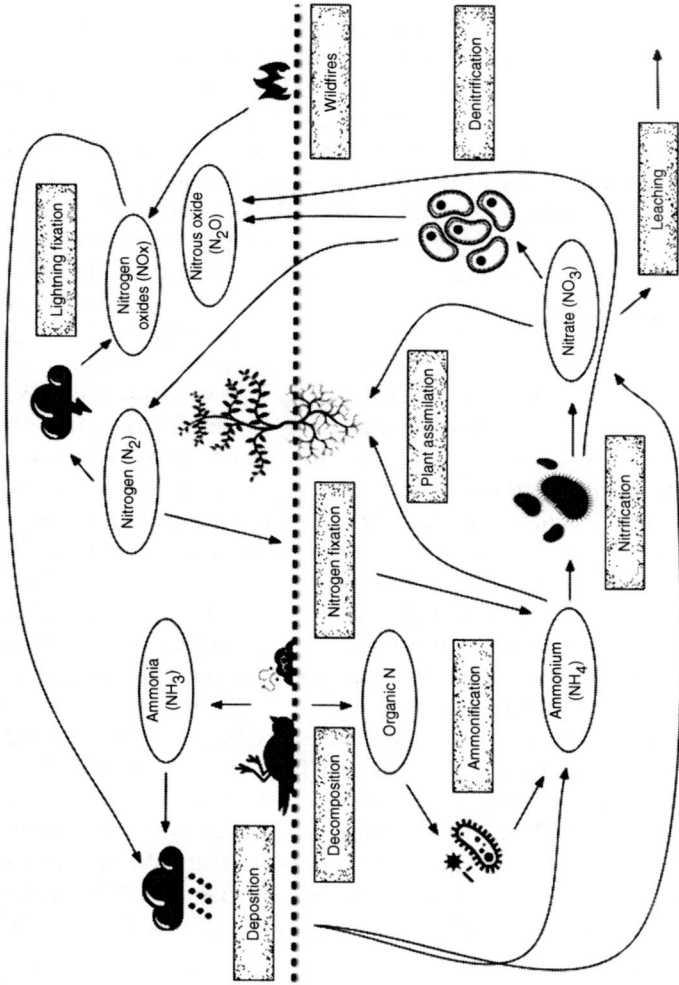

Figure 1.1 The natural terrestrial nitrogen cycle

The dotted line represents the land surface with the boxes showing the key processes by which nitrogen is cycled and the ovals showing the changing form of nitrogen as it undergoes these processes.

Source: Dave Reay

'fix' it and produce the reactive nitrogen all life on Earth requires[10]. With nitrogen fixation came a nutrient supply to hold together the lengthening global food chain, and with their ability to photosynthesise came the oxygen that would poison many of the earliest forms of life, yet power the evolutionary engine to breathtaking complexity.

Over the three billion or so years that cyanobacteria have been around, they have successfully survived in just about every home and lifestyle going. In the driest, the coldest, the most barren corners of the planet, look for them and they will be there. Build a new house and cyanobacteria will be first to move in. Splash in a newly formed puddle of rainwater and it is cyanobacteria that will swirl around your feet. As pioneers they are peerless, colonising everything from deserts and coral reefs to the shaggy coats of ponderous sloths and the boiling-hot springs of Yellowstone[11].

Millennia before either Scott or Amundsen first set foot on Antarctica, cyanobacteria had made the frozen continent their home. In its apparently lifeless Dry Valleys – a freezing desert of wind-blasted rocks under an ozone-thin atmosphere – cyanobacteria are everywhere[12]. This, most inhospitable of environments, hides its living secret within the rocks themselves. Beneath the porous rock surface, escaping the scouring desiccation outside, cyanobacteria carve out an existence. Here they receive enough light for photosynthesis while shaded from the DNA-bending doses of ultraviolet radiation on the surface. In the parched streambeds that snake between the rocks yet more cyanobacteria lie waiting, the first drops of moisture bringing them back to full nitrogen-fixing life. They are there even in the shifting snow and in ice-filled lakes. As Captain Scott and his men trudged across the Beardmore Glacier on their one-way trip to the South Pole, countless billions of cyanobacteria lay beneath their frostbitten feet.

They are incredible survivors. In cell-bursting saltpans, acid lakes and soils heavy with toxic metals, they thrive where other life-forms falter. In one test of their hardiness, cyanobacteria were cooled to a temperature of minus 196°C[13]. They still clung on to life. Unsurprisingly, when biologists look for evidence of alien life it is to the Mars-like desolation of Antarctica's Dry Valleys to which they turn for parallels[14]. If we are not alone in this universe, then cyanobacteria are top contenders as the littlest of little green men.

Pre-industrial nitrogen and humankind

If all plants could pull off the same nitrogen-fixing trick as cyanobacteria, then many of the famines and food shortages that have peppered human history may never have happened. Unfortunately, very few of them can. Staple food crops like wheat, rice and maize are actually very nitrogen-hungry and, with each successive harvest, a portion of the much-fought-over available nitrogen in the soil is removed. Keep this up for long enough and the soil's reserves become so depleted that the crops start to suffer, their growth becoming stunted and the harvests getting progressively smaller. When the early humans first started to swap a life of hunting and gathering for a more sedentary existence, they neatly side-stepped this problem by continually shifting their fields from place to place[15]. This allowed the crops to make use of a new stock of soil nitrogen each year and left the old fields to gradually recharge their stores. It was only when such wandering farming gave way to the sessile version familiar today that the need to somehow put nitrogen back into the soil came to the fore.

One obvious way of doing this was to recycle some of the nitrogen removed in the harvested crop. This took the form of either throwing the crop waste back onto the field, or collecting the manure – both animal and human – derived from the harvest and adding it back to the soil. Unfortunately, such recycling is a 'far from perfect' circle. Wastes like straw contain relatively low amounts of nitrogen, and it may be several months before they are broken down sufficiently to offer up the nitrogen held in their tissues. Manure can be more nitrogen-rich, but still represents only a small fraction of that which was originally in the harvested crop. At best then, such recycling can only delay the inevitable. Some source of new nitrogen is needed to keep a field churning out good harvests indefinitely, and it is here that the cyanobacteria-mimicking chemical magic possessed by beans and their wider plant family – the legumes – comes into its own.

What makes beans and other legumes like clover, peas and lentils so very special is their ability to take nitrogen gas out of the air and put it back into the soil in a reactive, readily usable form. For those neighbouring plants unable to pull off this soil-enriching alchemy, the extra nitrogen supply can provide a very welcome boost. Never ones to miss out on getting something for nothing, farmers have long been putting these living fertiliser factories to good use.

Precisely when farmers first cottoned onto the idea of using legumes to recharge their fields is hard to tell[16]. The remains of peas and lentils have turned up in excavations in Turkey dating back almost eight millennia, with the leftovers of a 6,000-year-old dinner of beans cropping up in a cave in Peru[17]. It seems the ancient Egyptians were also partial to legumes, both in this life and the next – pots of lentils have been found in their 4,000-year-old tombs.

So, legumes were adding extra nitrogen to the fields of some of the earliest human civilisations. But of course this may have simply been a little-noticed sideshow to the main event. Beans, peas and lentils make a nutritious harvest themselves and, as these crops ripen in late winter or early spring – well before cereal crops like wheat – they provide a source of food at otherwise very hungry times of the year. Such sustaining if rather bland benefits get nothing less than a biblical commendation. Genesis relates the story of Esau and Jacob, twin sons of Isaac who did not get on and who adopted very different approaches to life. Esau, the first born, was a man of the open country and a skilled hunter, while Jacob was more for the quiet life, staying at home, tending the crops and goats. Esau's skill as a hunter was apparently not infallible, as, one dinnertime, he arrived at Jacob's tent near starvation and implored his brother to give him some red-lentil stew. The animosity between the two brothers dated back to their jostling one another in the womb and so, with typical Old Testament aplomb, Jacob offered the stew only in return for his brother's birthright. Esau gave it up for the meal and left as the disgruntled loser to his newly empowered lentil-tending twin.

If, in the dawn of agriculture, the soil-enriching properties of legumes were sometimes secondary to ensuring a winter food supply and getting one over on your brother, by the Middle Ages a growing global population and the need to wring more and more food out of each field had made them priceless. As well as intercropping[18] – the practice of growing legumes amongst the normal crop – farmers around the world were now rotating their crops of rice, wheat and corn with a stand of beans or clover to keep the fields fertile.

Beans and peas were not only excellent at improving the soil while alive, their dead tissues were also nitrogen-rich[19] and so the whole crop was frequently ploughed into the soil to provide a luxuriant slice of green manure. Bean manure was a favourite from the start. In the 3rd century BC, the Greek philosopher Theophrastus was already extolling the virtues of bean manure, and by 1637 a Chinese scientist by the name of Sung Ying-hsing was advising rice growers to throw old soybeans into their fields to boost yields[20]. Every bean was apparently enough to fertilise about three square inches of rice paddy – truly magic beans.

By the 18th century, beans and the other legumes were playing a critical role in filling the stomachs of an increasingly populous and enlightened world. Some fields were now producing triple the amount of food compared to a few hundred years earlier. It was clear that the presence of legumes, dead or alive, in the soil made it more productive. But how exactly they did it, what that special something they provided to the soil was, remained a mystery. Only in this Age of Enlightenment, 10,000 years after farmers first starting growing legumes, did the many secrets of nitrogen begin to be unlocked. Tentatively at first and then with all the vigour and back-biting savagery that academia could muster, the race that Haber would eventually win had begun.

Enlightenment

For all its ubiquity, the many secrets of nitrogen have taken millennia to unlock. It was in the chill miasma of pollution and rarefied thinking that was 18th-century Edinburgh that its veil first began to be tweaked. Daniel Rutherford, a botany professor[21], found that the bulk of the air was made up of a single gas that was 'almost perfectly noxious' and which extinguished mice and flames alike. Across in Sweden a chemist called Carl Scheele had managed to identify two very different ingredients of the air, one very reactive, which he named 'fire air', the other stubbornly unreactive and dismissed as 'spoiled air'. Soon after, the former gas – and clearly the more exciting – was named Oxygene[22] by the brilliant French chemist Antoine Lavoisier[23]. The gas left behind after all the 'oxygene' had been used up remained nameless for more than a decade. In an age of rapid discovery and eye-catching experiments, it is perhaps unsurprising that nitrogen remained the ignored old spinster to the engaging debutante that was oxygen. Nitrogen didn't smell, it had no colour, you couldn't make it explode or even set fire to it. For scientists reaching to unlock the mysteries of life, the universe and everything, a gas whose main claim to fame was that it killed mice didn't really cut it.

Only in 1789 did Lavoisier finally get around to bestowing a name on this rather unprepossessing component of the air. He called it azote[22] – 'without life' – based on its ability to kill any animals forced to breathe it. The following year an alternative name for the gas was suggested – nitrogene – and, in much of the world, this new name stuck. The French still use azote though, a fitting legacy given that five years after he bestowed this name Lavoisier was guillotined.

Despite its damning classroom label as a periodic bore, nitrogen gas sports the twin mask of laughter and death as surely as any of its more reactionary forms. The reason the mice died in Daniel Rutherford's pioneering experiments was not that the nitrogen itself was poisonous, but that there was no oxygen available and so they quickly ended up 'without life' (Box 1.1). When the amount of nitrogen in the air we breathe increases and the oxygen levels drop away, the first warning signs are rapid breathing and a racing pulse. As they fall still lower, feelings of fatigue and nausea set in. At below 10 per cent, movement is impossible and loss of consciousness, coma and death ensue[24].

Box 1.1

Nitrogen asphyxiation

In the US, between 1992 and 2002, some 85 nitrogen asphyxiation incidents were reported. Eighty people were killed and 50 injured, with around 10 per cent of the lives lost being those of would-be rescuers overcome in confined spaces by the same suffocating conditions that had downed their colleagues. Many of these accidents are the result of a simple mix up of air supply tanks with nitrogen cylinders. A report from the US Chemical Safety and Hazard Board tells the tragic tale of just such a mix-up at a US nursing home[25]. Several of the home's residents suffered from respiratory diseases and needed regular supplies of bottled oxygen. Along with the standard oxygen cylinder delivery, the supplier one day delivered a bottle of pure nitrogen. This nitrogen cylinder was duly hooked up to the oxygen supply system and, as the regulator was opened, nitrogen gas flushed through the nursing home's network of oxygen lines. Four of the residents died and six more were injured.

By the start of the 19th century it was widely accepted that not just the air, but every living thing contained nitrogen. The idea that a good supply might be important to the growth of plants was now starting to gather momentum, though where this supply actually came from was to become an area of bitter contention. One of the greatest academics of the day, Baron Justus von Liebig[26], was the first to wade into the fray. He knew that the nitrogen in the manure that farmers added to their fields could not by itself provide enough for continually high yields, and that there must be another source. He reasoned that because the bulk of the nitrogen in the air was stubbornly unreactive, it must be the frequent drenching of ammonia – a readily usable form of nitrogen – in rain and snowfall that provided all the nitrogen the plants needed. Whenever rainwater was tested for ammonia it was always present, so Liebig seemed to be on the right track.

Having ticked nitrogen off his list, he started looking at the other minerals that plants needed for growth, ones that were not supplied in rainfall and so would be the most important as fertilisers to boost crop yields. He identified potash, lime and silica as key missing ingredients to a super-productive soil, predicting that the fields of the future would be prescribed such mineral boosters just as medicines are prescribed to ill patients. So confident was he that he had cracked the recipe for the perfect fertiliser that, in 1845, he started selling special mixtures made from bones, glass and ashes that, according to his calculations, would green the increasingly exhausted fields of England, but they did not work. Showing impressive courage in contradicting the statements of one of the world's leading scientists, John Lawes[27], a landowner, systematically tested fertilisers with and without nitrogen on the wheat fields of his family estate at Broadbalk in Hertfordshire, England. He and his colleague Joseph Gilbert – an old student of Liebig's – showed that wheat fields receiving fertiliser with added nitrogen grew much better than those fed with the nitrogen-free wonder minerals prescribed by Liebig. The great man was not best pleased, accusing Lawes and his colleagues of being 'a set of swindlers'. Yet the facts were there for all to see. Liebig's fertiliser enterprise folded within just three years of starting, the final boot to his theory being on the foot of a French chemist by the name of Jean-Baptiste Boussingault[28].

At the same time that Liebig had been dismissing manure as an insignificant source of nitrogen for plants, Boussingault was busily testing every potential fertiliser from dried blood to rotten cod to see how much nitrogen it contained (hair and wool emerged as the richest sources)[29]. He concluded that the value of any given fertiliser to a crop depended on how much nitrogen it contained. Now, with Leibig reeling from the red-faced truth that his mineral fertilisers failed to do what he said they would, Boussingault showed that the amounts of ammonia contained in rainfall were actually far lower than Liebig had assumed. Nitrogen was firmly back on centre stage.

Giggling chemists

Years before this spat kicked off, Boussingault had already shown that legumes could somehow increase the amount of available nitrogen in the soil[30]. Using pots of washed sand, he showed that clover and peas could increase the available nitrogen, but wheat and oats could not. What must have been incredibly frustrating for him and the many other scientists trying to establish how legumes pulled off this nitrogen-enriching trick was that they did not always do it. Back in the patchwork of the green and not-so-green fields of Broadbalk, the soils that had legumes grown in them showed huge increases in nitrogen. But when these same plants were grown under laboratory conditions in sterile soil they showed no sign of the nitrogen-adding abilities so apparent in the fields just outside. It was this annoyingly unpredictable appearance of extra nitrogen that ultimately revealed the legumes' trick.

Boussingault and Lawes were old men, Liebig long in his grave, when an elegant set of experiments finally showed that, while the legumes failed to thrive in pots of sterilised soil, once some fresh field soil was introduced their nitrogen-enriching abilities returned. The secret, then, lay in their special relationship with soil microbes. Kill off the bacteria by sterilising the soil and nitrogen enrichment by the legumes would stutter, but introduce a heady brew of bacteria-rich fresh field soil and the legumes could quickly get up to full soil-enriching speed. The two scientists responsible for uncovering this world-changing partnership – Hermann Hellriegel and Hermann Wilfarth – even identified exactly where in the plants this happened[31]. It was the small nodules on the roots of legumes, previously thought to be some kind of storage organ, that acted as the engine rooms in which plant and bacteria combined to 'fix' nitrogen.

Soon after the two Hermanns had published their work, the first species of nitrogen-fixing bacteria, Rhizobium[32], was isolated from pea root nodules, and around the world the search for similarly blessed bacteria intensified. As well as more species living in partnership with legumes, other bacteria – like cyanobacteria – were found that could live independently and still pull off the nitrogen-fixing trick. In soils, in plants, even in the oceans, the nitrogen-fixers were actively enriching the Earth. At the heart of this ability to fix nitrogen is an incredible enzyme called nitrogenase. This sludgy brown world-changing substance is able, at ordinary temperatures and pressures, to break apart the powerfully bonded nitrogen atoms in the air and fix them as reactive nitrogen. It is the kind of reaction chemists giggle about in their dreams.

For every molecule of nitrogen gas, nitrogenase is able to produce two of ammonia. To do this, however, it needs large amounts of energy. The global energy currency of life on Earth is called adenosine triphosphate – ATP for short – and a total of 16 ATPs are needed for the conversion of just one molecule of nitrogen[33]. Even then, it is hard work. The nitrogenase enzyme and the nitrogen molecule it is trying to break apart have to combine and separate no less than eight times until, after more than a second – a veritable age in terms of chemical reactions – the nitrogen is finally torn apart and reassembled as ammonia. All this is a huge drain; the valuable ATP could otherwise be used for growth and reproduction. Any organism that is to fix nitrogen successfully needs a bountiful source of energy to drive the whole process. For most of the nitrogen-fixers this source is, directly or indirectly, the Sun.

The bacteria nestled in the root nodules of legumes take part in a careful trade between the fatty acids produced by the plant during photosynthesis – and supplied to the bacteria to help power nitrogen fixation – and ammonia produced by the bacteria and transferred back to the plant. It seems an ideal partnership. These nitrogen-fixing bacteria are present in nearly every soil, waiting for the right type of seed to fall nearby and begin to grow. If the plant is the right one, is close enough and is at about the right stage – usually just before the leaves come out – then the rhizobia have their chance. The plant is not coy about inviting the rhizobia in. To let all the neighbouring rhizobia of the right type know that its roots are open for investigation, the plants release an attractant chemical into the soil. On picking up this welcome message the bacteria reply with their own chemical signal to the plant, this time inducing the nearby root hairs of the plant to curl up to make a pocket through which the bacteria can get into the root itself. Once in, the bacteria multiply fast, while around them the plant builds the walls of new tissue that form the nodule. In under a fortnight the plant can have a fully functioning nitrogen-fixing nodule and the bacteria a safe new home with a plentiful supply of food. Sometimes though, things do not quite work to plan. If there is already plenty of usable nitrogen around in the soil, the plant can break off relations with the rhizobia and leave them sitting unwanted in the soil while it gets all the nitrogen it needs directly from the soil. Where the marriage of plant and bacteria does go ahead, the relationship can be less than equal. Some of the rhizobia are happy to colonise the plant roots and make use of the food supplied by the plant, but give little or no usable nitrogen in return. Some plants have a way of dealing with such cheats. If a nodule is not pulling its weight in terms of supplying usable nitrogen, the plants appear able to cut off its supplies of oxygen and gradually reduce its size, redirecting resources to more productive nodules.

For the nitrogen-fixers that live independently of plants, getting enough energy by way of fatty acids to power nitrogen fixation becomes more difficult. Deprived of the plant's supply of fatty acids and nutrients, they must obtain such basic ingredients for the energy they require from elsewhere. By far the most successful of such free-living fixers are cyanobacteria. In a wonderful balancing act between conducting their own photosynthesis and nitrogen fixation, they are able to obtain the energy they need and then use this directly to power nitrogen fixation. This coupling has one very major drawback though: nitrogenase – that brown nitrogen-fixing wonder-enzyme – will not work when there is oxygen around, and of course photosynthesis produces loads of the stuff.

In the roots of legumes, the problem of too much oxygen crippling nitrogen fixation is neatly dealt with by the plant itself. The thick tissues of the nodules protect the bacteria from the oxygen-rich air outside, and a specialised oxygen carrier – similar to the haemoglobin in human blood and called leghaemoglobin[34] – is used to supply the nodules with just the right amounts of oxygen to keep the bacteria happy. Leghaemoglobin is bright red in colour and so gives the nodules of the legumes a pink colour. It binds so tightly to oxygen that it is able to mop up any nitrogenase-endangering amounts and reluctantly bleeds the small amount of oxygen the nitrogen-fixing bacteria need to the nodules. For those nodules not producing enough usable nitrogen for the plant, the shutting off of their oxygen supply can be seen by these nodules fading to white.

Out in the oxygen-rich soils and oceans, the free-living nitrogen-fixers have come up with a whole host of ways to keep their nitrogenase out of the way. In the sea some nitrogen-fixers are able to clump together, the ones on the outside of the ball capturing the sun's energy for photosynthesis while those in the centre, and so shielded from oxygen, switch to nitrogen fixation. Other bacteria invest even more energy in keeping oxygen at bay in the quest for reactive nitrogen, by ramping up respiration to use up nearby oxygen and protect their nitrogenase; some use as many as 30 molecules of ATP to fix a molecule of nitrogen, hugely expensive in terms of energy costs, but apparently worth it for the nitrogen it provides. Cyanobacteria have the trickiest balance to strike, with photosynthesis on the one hand generating new oxygen, and nitrogen fixation on the other hand requiring very low oxygen conditions. They do this using compartments called heterocysts[35]. By dividing up their cells into ones that are photosynthesising and others separated off from the high oxygen conditions, cyanobacteria are able to keep the nitrogen fixation going while at the same time garnering the energy they need to power it. There are also some that do not have these heterocysts and instead switch-over from photosynthesis in the light to nitrogen fixation when it is dark.

Nitrogen cycles

With fossil remains of cyanobacteria dating back more than three billion years, one might assume that, along with all the other nitrogen-fixers, their activities would have long ago ensured that the world's plants and animals were bathed in all the available nitrogen they would ever need. In reality, the enriching flow they provide is continually cycled back into the giant atmospheric pool through the actions of a group of microorganisms called the denitrifiers[36]. This group, made up of a whole host of bacteria and fungi, uses the 'fixed' nitrogen as a source of energy and in the process converts it back to nitrogen gas.

In terms of unlocking such secrets, the final decade of the 19th century was simply stunning. Within a few years of the 'Hermann & Hermann' legume breakthrough the full cycle of nitrogen – from atmosphere through soil and water into plants and animals and eventually back to the atmosphere – was revealed. The ammonia provided by the nitrogen-fixers was shown to be converted by nitrifying bacteria into even more readily available forms called nitrite and nitrate[37]. The denitrifiers then either got to work on this straight away, returning it to the atmosphere as nitrogen gas, or had to wait until whatever plant or animal they ended up in died and was broken down. They could then have another stab at getting hold of the nitrogen that remained (Box 1.2). Either by finally intercepting nitrogen recycled through many generations of plants and animals, or by being first at the table for shiny new nitrogen deposited via rain or snow, eventually the denitrifiers would get their chance.

Box 1.2

Death rights

Whether in the form of kings, prime ministers or Fritz Haber himself, humans are just another set of short-lived vectors in the global cycle of nitrogen. Break down the average human body into its constituent parts and two-thirds is water, about a sixth is fats, carbohydrates and minerals, and the rest is made up of the nitrogen-rich proteins and amino acids so precious to all life on Earth. This form of nitrogen is the building block for all DNA and RNA, and when humans die other organisms are not slow to take advantage.

Immediately following death, it is the bacteria and fungi already present in human intestines and the respiratory tract that begin the whole-body invasion. Any available oxygen is quickly used up as these microbes get to work on the easy-to-break-down tissues, the bacteria that prefer low-oxygen conditions quickly becoming dominant. New microbial species may migrate in from the soil or air, joining the increasingly fierce competition for the human-shaped pool of nitrogen-rich resources. In the soft tissues the cells begins to self-destruct – a process called autolysis, where enzymes begin to break them down from the inside. Within two to three days the microbes have responded to this cellular breakdown and putrefaction sets in. The speed at which all this happens depends on how warm it is, how much water is available and exactly which types of bacteria and fungi are present. The microbes first attack the proteins, breaking them down into more usable amino acids. As decomposition continues, carbon dioxide, hydrogen sulphide and methane are released, with the further breakdown of the amino acids releasing the aptly named and wretched-smelling gases putrescine and cadaverine.

(Continued)

Box 1.2 (Continued)

Some body tissues give up their nutritional treasures to this microbial onslaught more easily than others. The protein incorporated into the nervous system and in the linings of body cavities is the first to go. The protein in the form of collagen in the skin, in connective tissue and in muscles takes longer to break down, with keratin – the fibrous protein in hair and nails – being especially resistant to the tide of protein-splitting enzymes unleashed by the microbes. So stubbornly recalcitrant is keratin that otherwise completely skeletonised bodies may still boast a crop of hair. As decomposition of the body progresses and more and more of the proteins that hold it all together are broken down, the tissues and organs become one liquefied soup. Ultimately the flesh and organs are stripped away to leave a skeleton surrounded by a putrefying mass. This final stage of body breakdown can be lengthy. The bones are slowly eroded as the structural proteins within them are converted into peptides and amino acids. Eventually, both bones and hair are broken down, the human body and its life-giving load of reactive nitrogen entirely recycled.

Death and decomposition are the great levellers of life on Earth. Whether in the shape of a bull elephant or the parasitic worm in its gut, all life is part of a continuum of flowing resources embodied by whatever genetic variations are most successful at any given time. There is some irony in the way that Earth's current top consumers have found to indefinitely halt decomposition. Humans are now preserving their bodies in vats of liquid nitrogen.

Because the dinitrogen in gas or liquid form is so unreactive and does not freeze until it reaches minus 210°C, it is ideal for use in the preservation of a wide range of plant and animal material. From food to blood, from eggs and sperm to whole human bodies, it is often the preservative of choice[38]. Humankind has been attempting to halt the process of decomposition for many thousands of years. Embalming – the process of injecting alcohol into tissues – is just one of several methods designed to stave off microbial attack. The curing of meat can itself rely on denitrifying bacteria that produce such high levels of toxic nitrite in the tissues that other microorganisms are destroyed on contact. Salting, sugaring, pickling, all have the same microbe-bursting aim. More than 2,000 years after the preservation of Tutankhamen, humans are still playing the immortal body game.

Despite the rush of nitrogen-heavy Eureka moments at the end of the 19th century, the real prize – a way to control this newly exposed cycle – remained out of reach. Better recycling of manure could keep more of it in circulation; beans and other legumes could provide a fresh injection and help to bump up yields. But it just was not enough.

The population of Europe had now doubled to over 400 million, with North America seeing an explosion from seven million people at the start of the 19th century to more than 10 times that number by the end of the century. Soil exhaustion – the running down of the available nitrogen in farmland soils – was becoming an increasing problem. In the US, newly established fields that initially bore record-breaking harvests were starting to fail. If the many millions of hungry new mouths on either side of the Atlantic were to be fed, a new source of reactive nitrogen had to be found.

References

1. Witschi, H. Fritz Haber: December 9, 1868–January 29, 1934. *Toxicology* **149**, 3–15 (2000).
2. Erisman, J. W., Sutton, M. A., Galloway, J., Klimont, Z. & Winiwarter, W. How a century of ammonia synthesis changed the world. *Nature Geoscience* 1, 636–639, doi:10.1038/ngeo325 (2008).
3. Paerl, H. W. & Scott, J. T. Throwing fuel on the fire: synergistic effects of excessive nitrogen inputs and global warming on harmful algal blooms. *Environmental Science & Technology* **44**, 7756–7758, doi:10.1021/es102665e (2010).
4. Li, D. et al. A review of nitrous oxide mitigation by farm nitrogen management in temperate grassland-based agriculture. *Journal of Environmental Management* **128**, 893–903, doi:10.1016/j.jenvman.2013.06.026 (2013).
5. Bradley, M. J. & Jones, B. M. Reducing global NOx emissions: developing advanced energy and transportation technologies. *Ambio* **31**, 141–149 (2002).
6. Manchester, K. L. Man of destiny: the life and work of Fritz Haber. *Endeavour* 26, 64–69 (2002).
7. Ferguson, E. E. & Libby, W. F. Mechanism for the fixation of nitrogen by lightning. *Nature* **229**, 37, doi:10.1038/229037a0 (1971).
8. Tuck, A. Production of nitrogen oxides by lightning discharges. *Quarterly Journal of the Royal Meteorological Society* **102**, 749–755 (1976).
9. Navarro-Gonzalez, R., McKay, C. P. & Mvondo, D. N. A possible nitrogen crisis for Archaean life due to reduced nitrogen fixation by lightning. *Nature* **412**, 61–64, doi:10.1038/35083537 (2001).
10. Berman-Frank, I., Lundgren, P. & Falkowski, P. Nitrogen fixation and photosynthetic oxygen evolution in cyanobacteria. *Research in Microbiology* **154**, 157–164, doi:10.1016/S0923-2508(03)00029-9 (2003).
11. Miyamoto, K., Hallenbeck, P. C. & Benemann, J. R. Nitrogen fixation by thermophilic blue-green algae (cyanobacteria): temperature characteristics and

potential use in biophotolysis. *Applied and Environmental Microbiology* **37**, 454–458 (1979).

12. Pandey, K. D. et al. Cyanobacteria in Antarctica: ecology, physiology and cold adaptation. *Cellular and Molecular Biology* **50**, 575–584 (2004).

13. Nunez-Vazquez, E. J., Tovar-Ramirez, D., Heredia-Tapia, A. & Ochoa, J. L. Freeze survival of the cyanobacteria *Microcoleus chthonoplastes* without cryoprotector. *Journal of Environmental Biology/Academy of Environmental Biology, India* **32**, 407–412 (2011).

14. Andersen, D. T., McKay, C. P., Wharton, R. A. & Rummel, J. D. Testing a Mars science outpost in the Antarctic dry valleys. *Advances in Space Research: The Official Journal of the Committee on Space Research* **12**, 205–209 (1992).

15. Spengler, R. et al. Early agriculture and crop transmission among Bronze Age mobile pastoralists of Central Eurasia. *Proceedings. Biological Sciences/The Royal Society* **281**, 20133382, doi:10.1098/rspb.2013.3382 (2014).

16. Zohary, D. & Hopf, M. Domestication of pulses in the Old World: legumes were companions of wheat and barley when agriculture began in the Near East. *Science* **182**, 887–894, doi:10.1126/science.182.4115.887 (1973).

17. Phillips, R. Starchy legumes in human nutrition, health and culture. *Plant Foods for Human Nutrition* **44**, 195–211 (1993).

18. Ofori, F. & Stern, W. Cereal-legume intercropping systems. *Advances in Agronomy* **41**, 41–90 (1987).

19. Liebman, M., Graef, R. L., Nettleton, D. & Cambardella, C. A. Use of legume green manures as nitrogen sources for corn production. *Renewable Agriculture and Food Systems* **27**, 180–191, doi:10.1017/s1742170511000299 (2012).

20. Scorgie, M. E. & Ji, X.-D. Production planning in seventeenth century China. *Accounting History* **1**, 37–54 (1996).

21. Rutherford, D. *Dissertatio inauguralis de aere fixo dicto, aut mephitico: quam annuente summo numine: ex auctoritate reverendi admodum viri, Gulielmi Robertson, S.S.T.P. Academiae Edinburgenae Praefecti: nec non amplissimi senatus academici consensu, et nobilissimae facultatis medicae decreto: pro gradu doctoratus, summisque in medicina honoribus et privilegiis rite et legitime consequendis, eruditorum examini subjicit* (Edinburgi: Apud Balfour et Smellie, academiae typographos, M, DCC, LXXII, 1772).

22. Claude, G. & Cottrell, H. E. P. xxv, 418 pages (J. & A. Churchill, London, 1913).

23. Donovan, A. *Antoine Lavoisier: science, administration, and revolution.* (Blackwell, 1993).

24. Peers, C., Haddad, G. G., Chandel, N. S. & New York Academy of Sciences. In *Annals of the New York Academy of Sciences* **1177**, vi, 205 pp. (New York Academy of Sciences; John Wiley distributor, New York, Chichester, 2009).

25. Amyotte, P. R., MacDonald, D. K. & Khan, F. I. An analysis of CSB investigation reports concerning the hierarchy of controls. *Process Safety Progress* **30**, 261–265 (2011).

26. Brock, W. H. *Justus von Liebig: the chemical gatekeeper.* (Cambridge University Press, 1997).

27. Bawden, F. C. John Bennet Lawes (December 28, 1814–August 31, 1900) Joseph Henry Gilbert (August 1, 1817–December 23, 1901). Biographical sketches. *The Journal of Nutrition* **90**, 3–12 (1966).

28. Boussingault, J. B. *Rural economy, in its relations with chemistry, physics, and meteorology, or, An application of the principles of chemistry and physiology to the details of practical farming.* (H. Bailliere, 1845).

29. Aulie, R. P. Boussingault and the nitrogen cycle. *Proceedings of the American Philosophical Society* **114**, 435–479 (1970).

30. Burns, R. C. & Hardy, R. W. Nitrogen fixation in bacteria and higher plants. *Molecular Biology, Biochemistry and Biophysics* **21**, 189 (Springer-Verlag, 1975).

31. Ashby, S. Some observations on the assimilation of atmospheric nitrogen by a free living soil organism – *Azotobacter chroococcum* of Beijerinck. *The Journal of Agricultural Science* **2**, 35–51 (1907).

32. Dilworth, M. J. & SpringerLink (Online service). In *Nitrogen Fixation 7*, xix, 402 pp. (Springer, Dordrecht, 2008).

33. Postgate, J. R. *The fundamentals of nitrogen fixation.* (CUP Archive, 1982).

34. Atkins, C. A., Shelp, B. J., Storer, P. J. & Pate, J. S. Nitrogen nutrition and the development of biochemical functions associated with nitrogen fixation and ammonia assimilation of nodules on cowpea seedlings. *Planta* **162**, 327–333, doi:10.1007/BF00396744 (1984).

35. Kumar, K., Mella-Herrera, R. A. & Golden, J. W. Cyanobacterial heterocysts. *Cold Spring Harbor Perspectives in Biology* **2**, a000315, doi:10.1101/cshperspect.a000315 (2010).

36. Payne, W. J. *Denitrification.* (Wiley, 1981).

37. Prosser, J. I. & Society for General Microbiology. *Nitrification.* (Published for the Society for General Microbiology by IRL, 1986).

38. Cockcroft, P. D. *The low temperature storage of bovine embryos in liquid nitrogen* (Centre for Tropical Veterinary Medicine, University of Edinburgh, 1981).

2
Nitrogen and the Anthropocene

For a 19th-century world shrinking with each new shipping route, bringing in extra nitrogen not just from the manure heap down the road, but also from stocks thousands of miles away was now possible. Nutrient-rich waters and abundant fish stocks off the coast of Peru[1] had supported generation after generation of seabirds. The by-product of this long-standing fishy feast was huge accumulations of nitrogen-loaded bird manure, called guano. In the low-rainfall conditions of the region it built up year on year, and by the latter half of the century these mounds of off-white gold had become a vital prop to global food production[2,3].

The supplies could not last forever. More than half a billion tonnes were being extracted annually, and by 1872 Peru's supplies had all but disappeared. Attention increasingly turned to Chile and their own deposits of a nitrogen-rich mineral called caliche[4], but here too supplies were limited and costs high. Caliche was particularly sought after, as it could not only turn unproductive fields into lush breadbaskets, but also provided the raw material for explosives. Any nation wanting to go to war therefore needed a ready and plentiful supply. Chile itself ended up fighting with its neighbours Bolivia and Peru to gain control of this fabulously lucrative resource.

In the US and Europe, a reliance on importing such extra nitrogen from overseas brought with it all the insecurities and opportunities for blackmail inherent in oil and gas supplies today. To break these shackles they needed an independent means to enrich their fields, an artificial process that could mimic nature and somehow turn some of the vast stocks of nitrogen in the atmosphere into the usable forms that now held their economies together. They needed another chemist.

Fritz Haber

As the clouds of war gathered in Europe at the dawn of the 20th century, Germany, a powerhouse of industrialisation and the world's biggest importer of nitrate-rich caliche, had more to worry about than most. With so little coastline and the naval might of Britain on its doorstep, Germany could quickly be cut off from the nitrogen supplies so vital to its food production. Several ways of trying to mimic legumes and create reactive forms of nitrogen like nitrate and ammonia had already been tried. Some could be produced as a by-product of burning coal and, in the lab, small amounts were formed when making cyanide[5]. Neither could really keep up with demand, and the energy requirements for large-scale production were massive.

It was this powerful and increasingly paranoid Germany that finally provided the solution. Fritz Haber, a 36-year-old chemist working in the southwestern city of Karlsruhe, began by looking at why previous efforts might have failed. He knew that by finding the exact conditions at which ammonia (a highly reactive form of nitrogen) decomposed into its component parts he could in theory reverse the whole thing, pushing the nitrogen and hydrogen back through the process to form ammonia.

Haber worked out a recipe of temperatures and catalysts that should be more effective than anything tried before and set about systematically testing it. In 1904 he made his first batch of ammonia – not much, but proof that it could be done[6]. For Haber, the nitrogen story might have ended there. He decided that the amounts his method produced were too small to be of any commercial interest and turned his attention elsewhere.

It was three years later that an embarrassingly public and bileful dismissal of his results by another scientist brought Haber back to the nitrogen problem. Building on earlier experiments, he now threw increased pressure into the mix. It worked. Constantly tweaking the recipe of temperatures, pressures and catalysts to yield more and more ammonia, he was finally ready for the first full demonstration of his new method. On 2 July 1909, in front of a group of eager industry observers, Haber set his new apparatus to work. Into one end of the series of bolted metal chambers and twisting pipes flowed pressurised hydrogen and nitrogen gases. Passing over hot catalysts and then on through a condensing cooler, the gases were finally brought together as liquid ammonia[7]. Haber had done it. For five self-righteous hours it flowed out of the system in a steady dribble. With populations growing and a world war in the offing (Box 2.1), it was a dribble that would quickly become a torrent.

Box 2.1

Explosive nitrogen

If everything had gone to plan, the German war effort would have needed little help from Haber. Sweeping rapidly through Belgium and the industrial heartland of France during the August of 1914, the onslaught of the large and well-equipped German army looked like it would crush all resistance before it. It was just 40 miles short of Paris, in the Battle of the Marne, that the tide was turned. The British and French forced the German army to stall, retreat and then dig in. What followed on the Western Front were three years of bloody, entrenched warfare, stretching from the English Channel to Switzerland and involving millions of men.

Trench warfare demanded new strategies to break enemy lines, and initially it was massed artillery that was used by both sides to try and demolish defences and carve out another few yards of blasted frontline. The amounts of explosives that were rained down on the trenches by either side were staggering. In the Battle of the Somme alone the Allies fired 100,000 shells a day into the German lines in an effort to obliterate them. The number of deaths and injuries resulting from such bombardments was enormous, shrapnel and explosives accounting for 80 per cent of all casualties on the Western Front.

The curtains of high explosive that swept the fields of Flanders had their origins back in 9th-century China. Oft cited as the inventors of gunpowder, the Chinese apparently discovered it during failed attempts at alchemy. A mixture of charcoal, sulphur and potassium nitrate produced the explosive black powder, and in short order it was being made into bombs and hurled at startled enemies. By the time Guy Fawkes and his co-conspirators were stacking barrels of the stuff under a 17th-century House of Lords, with the aim of lifting Charles the First a few hundred feet off his throne, gunpowder had become a central tool of warfare around the world.

Access to gunpowder was now a matter of national security, with the Seven Years War between France and Britain giving a taster of the rush for raw materials and chemical expertise that would later define the two world wars.

Impressive explosive it may have been, but the gagging plumes of smoke that gunpowder produced meant riflemen and cannon crews spent much of their time firing blind into a pall of battlefield murk.

(Continued)

Box 2.1 (Continued)

Its use reached a high-water mark during the Napoleonic Wars and then subsided as a new smoke-less powder made from nitrocellulose came on the scene. This was more powerful than its black powder predecessor and was to provide the propellant for the billions of bullets that streamed across the battlefields of the 20th century.

Alfred Nobel[9], he of the Nobel Prize that Haber would receive for his nitrogen work, was closely involved with the development of this new generation of explosives. As well as dynamite – explosive of choice for mining companies and Wile E. Coyotes – Nobel invented the safer and more malleable highly explosive gelignite, later used by the provisional Irish Republican Army. These increasingly powerful nitrogen-based explosives were changing the face of warfare. From more accurate and deadly rifle fire to more devastating TNT-filled artillery shells, reactive nitrogen was now primed to blow the world order of the 19th century to pieces.

Haber's brilliant discovery was able to provide a secure and reliable source of ammonia, and the pressure was on to increase output. Carl Bosch[7], an expert in industrial chemistry, was tasked with perfecting large-scale production. He quickly developed a system that could pump out over a tonne of ammonia in a day. In the summer of 1914 the Haber–Bosch process[8] was churning out over 20 tonnes per day, with ever bigger and better factories planned. Germany was at last breaking free from the shackles of imported nitrogen. The timing was perfect. War had come again to Europe.

With the predicted blockade of its ports by the Royal Navy falling like an axe on its war strategy, Germany's supply of imported nitrate from which to meet the huge demand for high explosives was threatened. The ammonia now flowing from the Haber–Bosch process was the answer to the German war ministry's prayers. Carl Bosch and his team rapidly extended industrial production to transform the crop-boosting ammonia into trench-shattering nitric acid[10].

Without Haber and Bosch, the Great War might have been over by 1916 – a lack of munitions squeezing the German army into retreat and eventual surrender. With the availability of nitrogen-based explosives, the First World War raged on for a further two years and claimed the lives of almost 10 million men. The double-edged nature of Haber's discovery, one that could put food in the mouths of the world's hungry yet was now instrumental in the suffering and death of millions, did not seem to vex the man himself. This fierce nationalist believed Germany was simply defending herself and now began looking for new ways to use his expertise to decisively defeat the enemy. As the trenches got deeper and the fortifications stronger, even the nitrogen-propelled storm of rockets, shells and mines could not easily dislodge the opposing forces. The Germans needed something that could sweep through the densest lines of barbed wire, overwhelm the reinforced trench walls and penetrate into the corners of the deepest bunkers. For Haber the solution was obvious. They needed a chemical weapon[7].

Now in charge of a large research team in Berlin, Haber began to develop chlorine gas. Though the pre-war Hague conventions had made chemical warfare illegal, he saw it as a means to end the war swiftly and so avoid more suffering in the trenches. The Germans were not alone in their illegal development of chemical weapons. On the other side of the front line the French chemist Victor Grignard was working on a deadly gas called phosgene for the Allies. In the end, the race to develop chemical weapons was yet another that Haber won[11].

In the early evening of 22 April 1915 Haber oversaw the release of thousands of canisters of chlorine just upwind from the allied trenches near Ypres in Belgium. The greenish gas drifted towards lines of unsuspecting French soldiers. Heavier than air, it pooled in hollows and seeped into the trenches, with horrific results. Attacking eyes, throats and lungs, the gas attack broke the French lines and caused 15,000 casualties, of which 5,000 died. In terms of a weapon for trench warfare Haber's chlorine gas had proved a great success. In subsequent weeks it was used several more times on the Western Front and plans were now laid to deploy it against the Russian armies in the east.

Haber may have been achieving great success on the battlefield, but all was not well at home. His wife and fellow chemist, Clara, was appalled at the chemical hell her husband was unleashing. Her entreaties to him to halt this work fell on deaf ears. The night before Haber was due to leave to oversee the deployment of his deadly gas on the Eastern Front, a distraught Clara shot herself through the heart with his service revolver. If Haber felt any guilt about this it did not seem to sway his resolve. He left for the Eastern Front as planned[7].

Haber's contribution to the First World War and beyond was immense. The number of malnourished lives his chemical discoveries have saved is as awe-inspiring as the stomach-churning devastation they have wreaked. By the end of the Great War the chemical weapons used by both sides had accounted for the lives of more than a million men. Decorated by the army for his efforts, the post-war Haber continued to work hard for the greater glory of Germany. Unable to continue his chemical weapons research, he and his team now turned their attention to developing more effective fumigants and insecticides to protect the nation's precious food stores. Haber went on to lead an ultimately fruitless expedition aimed at paying off Germany's huge war reparations bill by extracting gold from seawater. By 1927 he was a depressed and ill man. Suffering from chronic heart pain only relieved by taking liberal quantities of nitroglycerine, this savagely patriotic man was not even able to live out his days in Germany. With the rise of Nazism, the Jewish-born Haber was forced into exile in 1933. He died the following year in Switzerland.

Hitler was growing in power and a Second World War, once again nourished by the discoveries of Haber and Bosch, was brewing. For those Jewish-born members of Haber's family still living in Nazi Germany, his legacy was to be all too personal. One of the most effective pesticides that he and his team had developed was to become the agent of destruction for huge numbers of Jews in Nazi death camps. It was called Zyklon B[12].

Human footprints on the global nitrogen cycle

The global flood that followed Haber's first trickle of lab-made ammonia in 1909 was staggering. As Britain, France and the US divided up the spoils of victory after the First World War, they employed Germany's expertise in making reactive nitrogen to develop new plants in their own countries. Britain had already chosen a small town called Billingham in northeast England as its production hub.[13] By the late 1920s the town's rapidly growing population, my own grandparents among them, were already living in the shadow of a seething industrial ammonia plant so extensive that it inspired Aldous Huxley to pen his *Brave New World*.

During the Second World War this and other plants were again turned to making explosives but, as the guns fell silent and the troops returned home, food and using reactive nitrogen to grow much more of it became the priority. Striding from the bloody confusion of the 1940s the US drove ahead with a huge expansion of nitrogen production. New plants sprang up wherever a good supply of natural gas and water to feed the Haber–Bosch process could be found. By the end of the 20th century, worldwide ammonia production was topping 80 million tonnes a year, almost all of this coming courtesy of the Haber–Bosch process. It had become an integral part of making everything from nylon and plexiglass to Lycra hot pants and rubber gloves. Most importantly, it had become the keystone of agricultural production around the world and a major weight behind humankind's deepening footprint on the global environment.

Today, more than a third of all land on Earth is farmed. To feed the fast-expanding human population of the 20th century, bumper-yield crop varieties requiring much larger nitrogen supplies were introduced, and year after year the farm soils of the world received increasingly rich injections of fertiliser to keep yields climbing. With little additional space available for expansion and a growing human population, the area available to feed each person has plummeted to just 2,500 square metres – half of what it was in the 1960s. Yet new crop varieties, better irrigation and large helpings of fertiliser have allowed more and more food to be squeezed out of these rapidly shrinking personal plots. This so-called Green Revolution[14] in agriculture could, and should, have made famine a thing of the past. For every extra kilogram of nitrogen fertiliser applied, the new varieties of rice, maize and wheat would produce more than double the amount of grain that the older varieties could muster. From the pre-Haber–Bosch days at the beginning of the century to the expanding and increasingly burger-hungry human population of more than six billion at its end, the average farm field saw its force-fed supply of reactive nitrogen increase by 10,000 per cent. Even more reactive nitrogen is required to continue to increase harvests and feed the still-expanding population of seven billion people on the planet today. As such, it is agriculture that now dominates our collective impact on the global nitrogen cycle and the myriad opportunities and threats to which it gives rise.

At the heart of all benefits or costs from agricultural nitrogen is the efficiency of its use. Nitrogen can take on many different forms as it passes along the food supply chain[15], where these flows are poorly managed, a domino effect of cascading negative impacts to ecosystems, human health and the climate can be set in motion.

A nitrogen cascade

The various types of nitrogen fertiliser added to fields around the world are delivered to the soil in a host of different ways. A shower of pellets or liquid applied to the surface, injections into the soil, or random hot spots of urine and manure dumped by contemplative livestock, each has the potential to boost the harvest and pay the farmer back in bumper yields. It is in the flow between crop sprayer or animal rear and the intended recipients that a host of leaks can spring (Figure 2.1).

First there are the losses to the air. Especially where the new nitrogen is sitting on the surface of the soil and plants, large amounts may be whipped away as ammonia gas[16] – the nose-busting form also common

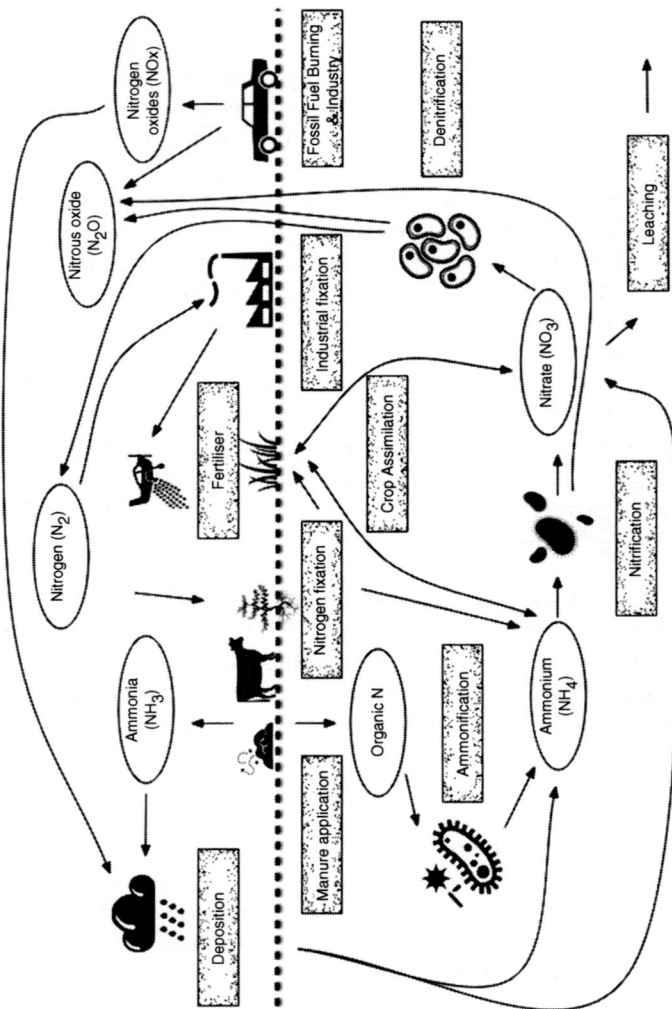

Figure 2.1   Human disturbance and the terrestrial nitrogen cycle

The dotted line represents the land surface with the boxes showing the key processes by which nitrogen is cycled and the ovals showing the changing form of nitrogen as it undergoes these processes.

Source: Dave Reay

to wet nappies. The fertiliser that the wind does not carry straight off then faces the massed ranks of plant roots and microbes, each intent on securing a slice of the bounty for themselves. One set of soil bacteria, called the nitrifiers[17,18], can transform the added nitrogen into nitrate, a form which is itself attacked by another group – the denitrifiers[18]. With each such transformation, yet more is converted to nitrogen gases and lost to the air. The plants do their best to capture the fertiliser intended for them, but invariably the amounts applied are far in excess of what they can utilise. With each passing hour, more and more escapes into the atmosphere, is locked up in microbial cells, or is converted to the highly mobile form called nitrate. Though crops can and do use this nitrate, it is often only fleetingly available. Once it enters the jumbled interconnected circuitry of water within the soil, it can easily slip down through the minute channels and pores and out of reach of the plant roots altogether[19].

For farmers, trying to plug these leaks is a continual battle. They can inject the fertiliser to stop the wind carrying so much away, but then the microbes are able to take a bigger share. They can add chemicals that will put a brake on the nitrifying bacteria[20], but this only works for a limited time. Despite their best efforts, the return farmers get in terms of their lashings of reactive nitrogen ending up in their harvest is miserably poor. In most cases, less than half is used directly by the plants for which it is intended. Some is stored away within the soil itself – a useful deposit account for future harvests, bound up in microbial cells or to soil particles[21]. Of the rest, the losses to the air above and the seepage waters below represent the first step in an enriching cascade that is threatening ecosystems and human health around the world.

The initial impacts of the plumes of ammonia that billow up from fields, barns and manure stores are largely invisible. Riding free on the wind, they join the millions of tonnes of extra reactive nitrogen that are belched into the air from fires, exhaust pipes and chimneys. These more direct injections into the atmosphere arise either from the nitrogen contained in the wood, coal or oil itself, or from the intense heat produced when such fuels are burnt that binds nitrogen gas together with oxygen. Whether courtesy of a patch of steaming cow pee or the exhaust pipe of a school bus, human activities add huge amounts of reactive nitrogen to the air and, sooner or later, it will return to Earth.

Downwind of the smoke stacks and cattle sheds the first enriching payloads deposited by air, rain and snow are greedily consumed by plants and bacteria alike, the ecosystems responding to this free meal with a flush of new growth. If the source is an especially potent one – like a battery chicken farm – the sheer eye-stinging intensity of the ammonia in the air can damage the vegetation[22]. In most cases though, the process is one of gradual saturation. As the years roll by, increasingly lush and fast-growing plants like grasses begin to take over, bullying any low-nitrogen specialists into submission. The shin-brushing beauty of most wildflower meadows is actually a product of the struggle going on below ground to get enough nitrogen. With scant resources, no one plant species can grow so quickly as to dominate the others and so results in a more diverse carpet of plants. A chronic rain of enrichment systematically destroys this fine balance. In woodlands and moorlands too, many of the native plants fall away to be replaced by tall stands of grasses rooted into a soil becoming increasingly overloaded[23].

They may be a matter of a few metres from the source or several hundred miles distant, but the fields and forests that receive these showers of reactive nitrogen are intricately linked with a much wider area through the waters that flow over and beneath the land surface. High in the hills above the breadbasket farmlands and smog-shrouded cities, the parcels of airborne fertiliser that arrive can start a domino effect that will snake back down to the plains below, and onwards to the sea.

The first growth-enhancing inputs might initially be used to build the tissues of a biodiversity-smothering grass. When this plant and the lush swards sprouting up around it die and break down, the multipurpose molecules of nitrogen they contain are free to rejoin the swelling nutrient pool. This pool can power the growth of yet more plants, leak extra nitrogen back into the atmosphere or, as the soils become overloaded, escape in ever-greater amounts to the waters that drain from the soil.

As the capacity of the soils and plants to use the aerial bombardments of nitrogen becomes saturated[24], losses to myriad small drainage channels begin to grow. Down through the hills these waters tumble, mixing with countless other streams similarly replete with the extra nitrogen dumped on the plants and soils that feed them. The volume of water grows, and with it the fertilising load it carries. The nitrate-heavy drainage waters of farm fields and rich effluents of sewage plants now pour into the flow, pulling a host of local pollution problems together into one super-charged river of enrichment. From flooding and blue babies to vomit-inducing clam chowder and suffocated fish, the worldwide impacts of these aquatic highways for humankind's wasted nitrogen are becoming increasingly stark. The nitrogen-induced greening of our lakes, rivers and seas brings with it a gamut of threats to ecosystems and human health.

This massive and ongoing perturbation of the global nitrogen cycle by humankind has strong parallels with that of the global greenhouse gas balance and our climate forcing. Like its carbon equivalent, humankind's footprint on the nitrogen cycle has grown from a minor impression caused by land-use change in the pre-industrial era to a world-changing stomp that has altered the stocks and flows of nitrogen everywhere, from the deepest ocean depths to the outer limits of our atmosphere. Just as fossil fuels have powered the engine of technological and economic development over the last century, so our mastery of reactive nitrogen production has sustained population growth unimaginable to our 19th-century predecessors. In the 21st century, these twin boons to human civilisation have become major interlinked threats to our way of life. We all now face the global challenge that is nitrogen and climate change.

References

1. Nixon, S. & Thomas, A. On the size of the Peru upwelling ecosystem. *Deep-Sea Research Part I – Oceanographic Research Paper* **48**, 2521–2528, doi:10.1016/s0967-0637(01)00023-1 (2001).
2. Mathew, W. M. Peru and the British Guano Market, 1840–1870. *The Economic History Review* **23**, 112–128 (1970).
3. Hollett, D. *More precious than gold: the story of the Peruvian guano trade.* (Associated University Press, 2008).
4. Urbansky, E. T., Brown, S. K., Magnuson, M. L. & Kelty, C. A. Perchlorate levels in samples of sodium nitrate fertilizer derived from Chilean caliche. *Environmental Pollution* **112**, 299–302 (2001).
5. Bartell, F. E. Nitrogen fixation by the cyanide process. *Journal of Industrial and Engineering Chemistry* **14**, 699–704, doi:10.1021/ie50152a008 (1922).
6. Haber, F. & van Oordt, G. On the formation of ammonia from the elements. *Zeitschrift für Anorganische Chemie* **44**, 341–378, doi:10.1002/zaac.19050440122 (1905).
7. Smil, V. *Enriching the earth: Fritz Haber, Carl Bosch, and the transformation of world food production.* (MIT Press, 2001).
8. Kandemir, T., Schuster, M. E., Senyshyn, A., Behrens, M. & Schlogl, R. The Haber-Bosch process revisited: on the real structure and stability of 'ammonia iron' under working conditions. *Angewandte Chemie* **52**, 12723–12726, doi:10.1002/anie.201305812 (2013).
9. Skagegård, L.-Å. *The remarkable story of Alfred Nobel and the Nobel Prize.* (Konsultförlaget, 1994).
10. Miles, F. D. *Nitric acid : manufacture and uses.* Imperial Chemical Industries. 75 pp. (Oxford University Press, London, 1961).
11. Johnson, J. A. Master mind: the rise and fall of Fritz Haber, the Nobel laureate who launched the age of chemical warfare. *Technology and Culture* **47**, 835–837, doi:10.1353/tech.2006.0231 (2006).
12. Bloch, P. Zyklon-B + Nazi extermination technology. *Esprit* **9**, 53–56 (1980).

13. Hunt, L. The ammonia oxidation process for nitric acid manufacture. *Platinum Metals Review* **2**, 129–134 (1958).
14. Khush, G. S. Green revolution: preparing for the 21st century. *Genome/ National Research Council Canada = Genome/Conseil national de recherches Canada* **42**, 646–655 (1999).
15. Galloway, J. N. The global nitrogen cycle: past, present and future. *Science in China. Series C, Life sciences/Chinese Academy of Sciences* **48 Spl issue**, 669–677 (2005).
16. Tian, G., Cai, Z., Cao, J. & Li, X. Factors affecting ammonia volatilisation from a rice-wheat rotation system. *Chemosphere* **42**, 123–129 (2001).
17. Prosser, J. I. & Society for General Microbiology. *Nitrification*. (Published for the Society for General Microbiology by IRL, 1986).
18. Payne, W. J. *Denitrification*. (Wiley, 1981).
19. Dunbabin, V., Diggle, A. & Rengel, Z. Is there an optimal root architecture for nitrate capture in leaching environments? *Plant, Cell & Environment* **26**, 835–844 (2003).
20. Dittert, K., Bol, R., King, R., Chadwick, D. & Hatch, D. Use of a novel nitrification inhibitor to reduce nitrous oxide emission from (15)N-labelled dairy slurry injected into soil. *Rapid Communications in Mass Spectrometry: RCM* **15**, 1291–1296, doi:10.1002/rcm.335 (2001).
21. Xu, G., Fan, X. & Miller, A. J. Plant nitrogen assimilation and use efficiency. *Annual Review of Plant Biology* **63**, 153–182, doi:10.1146/annurev-arplant-042811-105532 (2012).
22. van der Eerden, L. J. M., de Visser, P. H. B. & van Dijk, C. J. Risk of damage to crops in the direct neighbourhood of ammonia sources. *Environmental Pollution* **102**, 49–53, doi:10.1016/s0269-7491(98)80014-6 (1998).
23. Cleland, E. E. & Harpole, W. S. Nitrogen enrichment and plant communities. *Annals of the New York Academy of Sciences* **1195**, 46–61, doi:10.1111/j.1749-6632.2010.05458.x (2010).
24. Aber, J. D. Nitrogen cycling and nitrogen saturation in temperate forest ecosystems. *Trends in Ecology & Evolution* **7**, 220–224, doi:10.1016/0169-5347(92)90048-G (1992).

3
Nitrous Oxide as a Driver of Climate Change

> I have now discovered an air five or six times as good as common air . . . nothing I ever did has surprised me more, or is more satisfactory.
> (Joseph Priestley)

Nitrous oxide represents nitrogen's most direct driver of global climate change. Commonly known as laughing gas, this relatively inert gas was first described by Joseph Priestley in 1772 (Box 3.1), and it has since become a stock anaesthetic and analgesic for doctors, midwives and dentists everywhere[1]. Alongside the enlightenment boom in the use of nitrous oxide for medicine and recreation, its concentration in the atmosphere also began to rise. The source of this increase lay not in the labs of chemists or the drawing rooms of giggling London aristocrats, but in the rapid expansion of agriculture and industrialisation occurring around the world.

Until the 18th century the concentration of nitrous oxide in the atmosphere had remained at around 270 parts per billion (ppb) for several millennia[2]. This is a tiny amount – one part per billion being equivalent to a cup of water diluted into four Olympic-sized swimming pools – and hence nitrous oxide and its similarly rare greenhouse gas cousins carbon dioxide and methane are often called 'trace gases'.

From the ice core record we know that nitrous oxide concentrations went as low as 200 ppb in glacial periods and up to 280 ppb during warmer phases. These long-term swings were due to changes in nitrogen cycling around the planet, with warmer times unlocking more reactive nitrogen from frozen soils and waters.

Then, as the area of farmland increased to meet rising food demands in the 1700s, things began to change.

Box 3.1

Nitrous oxide in medicine

In 1772 the chemist Joseph Priestly was having a jolly time of things. Amongst other discoveries, he had isolated a gas he described as *phlogisticated nitrous air*. He noted that inhaling it induced a euphoric state of mind, but seemed to have no other side effects. This *phlogisticated nitrous air* or 'laughing gas' as it became commonly known was the powerful greenhouse gas nitrous oxide. Its happy effects were obviously a winner with Priestly, prompting him to describe it thus: 'I have now discovered an air five or six times as good as common air . . . nothing I ever did has surprised me more, or is more satisfactory.' By the 1790s, leading scientists, writers and socialites alike were regularly enjoying the effects of inhaling nitrous oxide, with the chemist Humphry Davy being a prime organiser of these 'laughing gas parties'. In addition to its recreational use, others quickly realised what huge benefits this gas could provide in terms of providing pain relief. One such man was Gardner Quincy Colton, a medical student in the US, who left college to tour the country demonstrating the pain-relieving powers of nitrous oxide. In 1844 Horace Wells, a dentist, visited this travelling circus, realised how effective this could be for use in his dental operations, and began to use it. By the 1860s Gardner Quincy Colton had given up touring and was himself using nitrous oxide on thousands of dental patients each year. Since the late 19th century, use of nitrous oxide as an anaesthetic has been common practice, with 'gas and air' (a 50:50 combination of nitrous oxide and oxygen) now frequently used for pain relief during childbirth and in emergency medicine.

The ploughing up of new land first allowed greater mineralization of soil organic nitrogen[3], thereby increasing the amount of reactive nitrogen available. Greater use of legume crops also meant more and more nitrogen was fixed from the atmosphere[4], and burgeoning fossil fuel combustion further ramped up the amounts of nitrogen deposited by wind, rain and snow[5]. This big increase in available nitrogen helped boost crop production, but it meant that nitrous oxide emissions began to creep upwards too, with the microbes most responsible for nitrous oxide production (the nitrifiers and denitrifiers) now responding to a strengthening anthropogenic nitrogen flow.

Nitrification and nitrous oxide

The form of nitrogen is crucial to its impact on our environment; some forms (such as nitrate) are very soluble in water and so often dominate water pollution issues, while others (such as ammonia) are very volatile and may represent major air pollution problems. It is during these conversions of one type of nitrogen to another (especially nitrification and denitrification) that nitrous oxide can also be produced.

Nitrification is the process whereby ammonia – a reduced form of reactive nitrogen – is oxidised to produce nitrate (an oxidised form)[6,7] (Figure 3.1). Both on land and in the oceans of the world, microorganisms are at the heart of this transformation. The bacteria responsible, called nitrifiers and usually belonging to the microbial groups nitrosomonas and nitrobacter, derive energy as each molecule of ammonia steps up through the oxidation pathway on its way to becoming nitrate. As its name suggests, the oxidation of ammonia requires oxygen and so is most commonly seen in well-drained soils or well-oxygenated surface waters. These microbes play a crucial role in the global nitrogen cycle and its many interactions with climate change, but it is when their ammonia-oxidising efforts go slightly off track that they have the most direct impact on global warming.

As the nitrifiers transform ammonia into nitrite in the first step of nitrification, some of the products produced in this process can decompose and form nitrous oxide before the step is fully completed. Still more

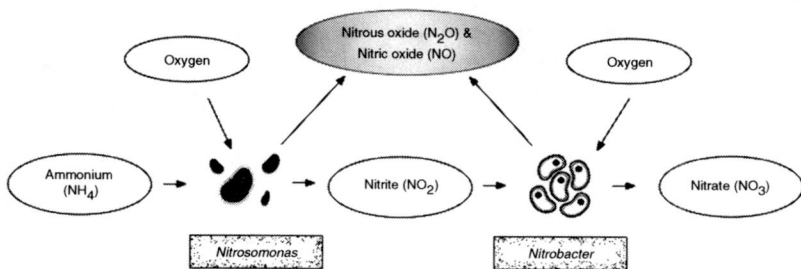

Figure 3.1   Nitrification

The boxes show the two groups of bacteria (nitrobacter and nitrosomonas) most often associated with nitrification. This process is usually aerobic (i.e. requires oxygen) with both nitrous oxide and nitric oxide being potential by-products. Anaerobic ammonium oxidation (called 'anammox') is also important, especially in aquatic environments, with the anammox bacteria able to convert the ammonium and nitrite directly to dinitrogen gas (N_2).

Source: Dave Reay

nitrous oxide may be produced even when the nitrite step is completed, with the nitrite being reduced to nitrous oxide and nitrogen gas instead of making the final onward step of oxidation to nitrate[8].

Denitrification and nitrous oxide

While the nitrifiers can be important producers of nitrous oxide in some environments, their contribution to climate change is often dwarfed by another group of microorganisms – the denitrifiers – whose actions are almost the complete opposite of their oxygen-loving namesakes[9]. Instead of gaining their energy from the oxidation of ammonium through to nitrite and nitrate, the denitrifiers start with nitrate and reduce it right back to nitrogen gas[7] (Figure 3.2).

To do this, denitrifiers need a low- or zero-oxygen environment, and so can thrive in any environments rich in nitrate but poor in oxygen, such as wet soils and sediments. There are three main steps that reduce nitrate down to nitrogen gas: the first transforms the nitrate to nitrite, the second turns the nitrite into nitrous oxide and nitric oxide, and the final step produces nitrogen gas. Under ideal conditions the denitrifiers do exactly this, with any available nitrate disappearing from the soil or water and being emitted in the relatively benign form of dinitrogen gas. It is when this transformation becomes imperfect that large emissions of nitrous oxide can begin[10].

If the oxygen levels are too high, the temperature too low or the conditions too acidic, then the crucial final step that turns nitrous oxide into dinitrogen may fail[11]. If such conditions are widespread and there is plenty of nitrate available, the denitrifiers can quickly become major sources of nitrous oxide to the atmosphere. In most soils, seas and sediments the microbial environment is a constantly changing patchwork of different conditions, sometimes ideal for the nitrifiers, sometimes ideal for the denitrifiers, and most often imperfect for either. As the

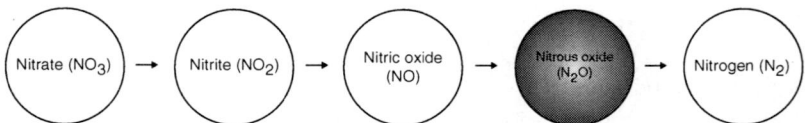

Figure 3.2 Denitrification

The circles show the various forms of nitrogen as it passes through the denitrification process. Nitrous oxide emissions occur when the process is incomplete – such as can occur where the environment is not fully anaerobic (oxygen-free).

Source: Dave Reay

nitrifiers are a great source of the nitrate the denitrifiers need, these microbes are often closely associated with one another. However, with one group being oxygen-loving and the other oxygen-hating, this so-called 'coupled nitrification-denitrification'[12] is often an unstable relationship that bears nitrous oxide as its climate-altering offspring.

Before major human interventions to the nitrogen cycle, the denitrifiers of the world had to rely on the sparse fare of nitrate derived from lightning strikes and whatever the nitrifiers provided via the natural cycling of reactive nitrogen. Along with the big human-induced increases in reactive nitrogen inputs into the land and oceans that started in the 18th century came a growing feast for nitrifiers and denitrifiers alike, and a super-charge to their emissions of nitrous oxide.

As food demand and fossil fuel burning further increased through the 18th and 19th centuries, so soil nitrous oxide emissions continued to grow[5]. By 1900 the concentration of nitrous oxide in the atmosphere had crept up to 280 ppb[2]. However, it was Fritz Haber's 1914 discovery of ammonia synthesis, and the subsequent explosion in the use of synthetic nitrogen fertilisers, that induced the surge in atmospheric concentrations that has continued to this day. Over the past few decades concentrations of nitrous oxide in the atmosphere have been increasing by around 0.25 per cent each year, with the global concentration having now reached 324 ppb – some 20 per cent higher than the pre-industrial average[2].

Climate forcing

Though its concentration is still very small relative to that of other major greenhouse gases such as carbon dioxide and methane, nitrous oxide has two attributes that make it a powerful greenhouse gas. First, it is chemically rather inert, which means that once it is emitted into the atmosphere it can remain there for a long time – the average molecule of nitrous oxide in our atmosphere has a lifetime of around 120 years[2]. Second, every nitrous oxide molecule is itself a very effective absorber and re-emitter of heat (infrared radiation)[13]. Infrared radiation is emitted from Earth at a range of wavelengths from 0.7 to 70 microns in size. Across much of this range the outgoing energy is absorbed and re-emitted by water vapour – the leading greenhouse gas on our planet – while at some wavelengths carbon dioxide and methane are the key greenhouse gases. It is at wavelengths of 4–5 and 7–8 microns that nitrous oxide becomes a big player (Box 3.2). In these narrow atmospheric windows that would otherwise allow heat radiation to escape from Earth, nitrous oxide effectively blocks the way – making it an especially powerful greenhouse gas.

Box 3.2

Measuring nitrous oxide

The measurement of nitrous oxide often relies on exactly the infra-red energy absorption characteristics that make it such a powerful greenhouse gas. Using a technique pioneered by the great 19th-century scientist John Tyndall[14], the air to be tested is passed over a heat source (emitting infrared radiation) and a detector then measures how much of the energy is absorbed at the precise wavelengths where nitrous oxide is known to be active. The long atmospheric lifetime of nitrous oxide means that it is very well mixed in our atmosphere, and so changes in global concentrations can be measured using just a few monitoring stations around the world. The downside of this long-lived and well-mixed nature is that it makes it difficult to pinpoint an exact source of emissions. To do that, a range of other methods are used to measure the emissions of nitrous oxide from specific areas, ranging from a few centimetres to many square kilometres. For small-scale (up to ~10 m²) measurements rigid chambers are usually used. These are placed on the surface of the soil or water, their lids sealed, and the nitrous oxide emissions measured by tracking the rate at which nitrous oxide concentrations in the chamber changes.[15] For larger areas (km²), nitrous oxide concentrations are measured at the top of towers or by aircraft, where the emissions can be assigned to a 'footprint' on the land or ocean based on the direction and speed of the wind that passes over the nitrous oxide sensor[16].

These properties of long life and strong infrared absorption mean that, over a 100-year time horizon, each tonne of nitrous oxide emitted has a global warming potential (GWP) equivalent to releasing 300 tonnes of carbon dioxide[13]. In terms of global radiative forcing – the extent to which something alters the radiation balance of the Earth – nitrous oxide's influence went up from a 0.1 watt per square metre ($W\ m^{-2}$) addition in 1980 to a 0.18 $W\ m^{-2}$ addition in 2011[2]. Nitrous oxide now makes up around eight per cent of all human-induced climate forcing due to greenhouse gas emissions. On current trends, its radiative forcing will exceed 0.5 $W\ m^{-2}$ by the end of this century, adding around 0.4°C to global temperatures[17]. As such, nitrous oxide represents the third most important contributor to human-induced climate change, having about one-tenth of the impact of carbon dioxide despite a concentration that is still 1,000 times lower.

Ozone hole-punch

When major depletion of the ozone layer was reported in the 1970s and 1980s, it was chlorofluorocarbons (CFCs) that were identified as the major driver of this loss[18]. Together with other ozone-depleting substances (ODSs) that contained chorine or bromine (called the halocarbons), the production, use and release of CFCs was quickly addressed through the introduction of the Montreal Protocol in 1987[19]. Since that time emissions of these major ODSs have been greatly reduced and there are now signs that the rate of ozone loss has slowed[20]. Worryingly though, one key ozone depleter was not included in the Montreal Protocol and its concentration in our atmosphere is growing by the year. Yet again, the culprit is nitrous oxide[21].

As we have already seen, in the lower atmosphere (called the troposphere) nitrous oxide is such a powerful greenhouse gas because it is so stable and lasts for many decades. It is when nitrous oxide molecules seep up into the upper atmosphere (the stratosphere) that they are prone to attack by a combination of sunlight and oxygen radicals. Most of the nitrous oxide is destroyed by sunlight, but where it instead reacts with oxygen radicals it forms nitrogen oxides, and it is these highly reactive substances that can then go on to deplete ozone.

Although only five to ten per cent of nitrous oxide in the stratosphere contributes to ozone depletion in this way, the much greater concentrations of nitrous oxide compared to other ODSs make it a major player in global ozone depletion[22]. When the Montreal Protocol was created to tackle ozone depletion the important role of nitrous oxide was already known, yet its emissions remained unregulated. As a result, the success story that is the reduction of CFCs under the Montreal Protocol has today pushed nitrous oxide up the ozone-depleter rankings to a point where it is responsible for more than 20 per cent of all ozone depletion[2]. With further reductions in CFCs combined with ever-increasing nitrous oxide emissions over the coming decades, the importance of nitrous oxide in terms of ozone depletion is set to grow and grow. By 2100 it is estimated that it will be causing more ozone depletion than all the halocarbon gases put together.

The ozone layer's primary role is to intercept short-wave radiation from the sun (ultraviolet rays in particular) before they reach Earth. Ozone depletion therefore means that more of this energy passes through our atmosphere and reaches the surface. This is a direct threat in terms of ultraviolet damage to plants, animals and humans[23], hence the urgency with which the Montreal Protocol was adopted. In terms of climate change, ozone depletion has two major effects that work in opposite directions. The extra energy received by the Earth's surface means that more heat energy (infrared) is then emitted, and so additional warming of the surface and lower atmosphere can occur. However, the reduction in ozone – which can also act as a greenhouse gas – in the stratosphere means that heat energy is also able to escape from the Earth more easily, leading to a cooling effect. Adding these two effects together gives a slight overall cooling effect from ozone depletion. In terms of nitrous oxide's full role as a driver of climate change since the industrial revolution, its role as a greenhouse gas in the troposphere has added about 0.18 W m^{-2}, while its role as an ozone depleter in the stratosphere has subtracted 0.01 W m^{-2}.

References

1. Rooks, J. P. Safety and risks of nitrous oxide labor analgesia: a review. *Journal of Midwifery & Women's Health* **56**, 557–565, doi:10.1111/J.1542–2011.2011.00122.X (2011).
2. Stocker, T. *Climate change 2013: the physical science basis: Working Group I contribution to the fifth assessment report of the Intergovernmental Panel on Climate Change.* (Cambridge University Press, 2014).
3. Davidson, E. A. The contribution of manure and fertilizer nitrogen to atmospheric nitrous oxide since 1860. *Nature Geoscience* **2**, 659–662, doi:10.1038/ngeo608 (2009).
4. Galloway, J. N. The global nitrogen cycle: past, present and future. *Science in China. Series C, Life sciences/Chinese Academy of Sciences* **48 Spl. issue**, 669–677 (2005).
5. Galloway, J. N. & Cowling, E. B. Reactive nitrogen and the world: 200 years of change. *Ambio* **31**, 64–71 (2002).
6. Prosser, J. I. & Society for General Microbiology. *Nitrification.* (Published for the Society for General Microbiology by IRL, 1986).
7. Baggs, E. & Philippot, L. Microbial terrestrial pathways to nitrous oxide. In *Nitrous Oxide and Climate Change*, edited by K. Smith, 4–35 (Earthscan, London, 2010).
8. Bremner, J. M. & Blackmer, A. M. Nitrous-oxide – emission from soils during nitrification of fertilizer nitrogen. *Science* **199**, 295–296, doi:10.1126/science.199.4326.295 (1978).
9. Payne, W. J. *Denitrification.* (Wiley, 1981).
10. Saggar, S. et al. Denitrification and N$_2$O:N$_2$ production in temperate grasslands: processes, measurements, modelling and mitigating negative impacts. *The Science of the Total Environment* **465**, 173–195, doi:10.1016/j.scitotenv.2012.11.050 (2013).
11. Pan, Y., Ye, L., Ni, B. J. & Yuan, Z. Effect of pH on N$_2$O reduction and accumulation during denitrification by methanol utilizing denitrifiers. *Water Research* **46**, 4832–4840, doi:10.1016/j.watres.2012.06.003 (2012).
12. Wunderlin, P., Mohn, J., Joss, A., Emmenegger, L. & Siegrist, H. Mechanisms of N$_2$O production in biological wastewater treatment under nitrifying and

denitrifying conditions. *Water Research* **46**, 1027–1037, doi:10.1016/j.watres. 2011.11.080 (2012).

13. Reay, D. S. et al. Nitrous oxide: importance, sources and sinks. In *Greenhouse Gas Sinks*, edited by D. Reay, C. N. Hewitt, K. Smith & J. Grace, 201–206 (CABI, Wallingford, UK, 2007).

14. Tyndall, J. *Contributions to molecular physics in the domain of radiant heat: a series of memoirs published in the 'Philosophical transactions' and 'Philosophical magazine', with additions*. (Longmans, Green, 1872).

15. Clayton, H., Arah, J. R. M. & Smith, K. A. Measurement of nitrous-oxide emissions from fertilized grassland using closed chambers. *Journal of Geophysical Research: Atmospheres* **99**, 16599–16607, doi:10.1029/94jd00218 (1994).

16. Jones, S. K. et al. Nitrous oxide emissions from managed grassland: a comparison of eddy covariance and static chamber measurements. *Atmospheric Measurement Techniques* **4**, 2179–2194, doi:10.5194/amt-4-2179-2011 (2011).

17. Kroeze, C. Nitrous-oxide and global warming. *Science of the Total Environment* **143**, 193–209, doi:10.1016/0048-9697(94)90457-x (1994).

18. Crutzen, P. J., Isaksen, I. S. A. & McAfee, J. R. Impact of chloro-carbon industry on ozone-layer. *Journal of Geophysical Research – Oceans and Atmospheres* **83**, 345–363, doi:10.1029/JC083iC01p00345 (1978).

19. United Nations Environment Programme. Ozone Secretariat. *Handbook for the Montreal Protocol on substances that deplete the ozone layer*. 3rd edn. (Ozone Secretariat, 1993).

20. Newchurch, M. J. et al. Evidence for slowdown in stratospheric ozone loss: first stage of ozone recovery. *Journal of Geophysical Research: Atmospheres* **108**, 13, doi:10.1029/2003jd003471 (2003).

21. Portmann, R. W., Daniel, J. S. & Ravishankara, A. R. Stratospheric ozone depletion due to nitrous oxide: influences of other gases. *Philosophical Transactions of the Royal Society B: Biological Sciences* **367**, 1256–1264, doi:10.1098/rstb.2011.0377 (2012).

22. Randeniya, L. K., Vohralik, P. F. & Plumb, I. C. Stratospheric ozone depletion at northern mid latitudes in the 21(st) century: the importance of future concentrations of greenhouse gases nitrous oxide and methane. *Geophysical Research Letters* **29**, 4, doi:10.1029/2001gl014295 (2002).

23. Rousseaux, M. C. et al. Ozone depletion and UVB radiation: impact on plant DNA damage in southern South America. *Proceedings of the National Academy of Sciences of the United States of America* **96**, 15310–15315 (1999).

4
Nitrous Oxide Sources

As human impacts on the global nitrogen cycle grew rapidly from the industrial revolution onwards, human-induced nitrous oxide emissions also increased. Today, so-called 'natural sources' still account for 60 per cent of worldwide emissions each year, but the human-induced component is becoming an ever-larger part of the global budget[1]. An estimated 16 million tonnes of nitrogen are emitted each year as nitrous oxide (nitrous oxide-N), with around 10–12 million tonnes coming from natural sources and a further 6–8 million tonnes arising from human activities (Table 4.1).

Nitrous oxide in prehistory

Before direct measurements began in the late 20th century, the most powerful tool available to estimate the nitrous oxide concentration in the atmosphere was the ice core record (Box 4.1). These ice cores are collected from ice sheets in the Arctic and Antarctic that can be more than two miles deep and hold records of the past atmosphere dating back more than a million years[6]. Because nitrous oxide lasts in the atmosphere more than a century, on average, the concentration shown in the ice core record can also be used to estimate global emissions at that time. The ice core record indicates that concentrations have remained relatively stable over most of the last 4,000 years[7,8], but over time-scales of hundreds of thousands of years they show more variation as the planet moves between cold (glacial) and warm (interglacial) periods; concentrations have increased by around 40 per cent following the end of the last glaciation[9].

Since the 18th century natural nitrous oxide emissions are thought to have remained fairly steady, at between 10 million and 12 million

Table 4.1   Estimates of global anthropogenic nitrous oxide emissions (Tg N y⁻¹) from different sources for the 1990s[2] and 2000[3]

Source	1990s	2000
Agriculture	2.8 (1.7–4.8)	3.8[a]
Human excreta	0.2 (0.1–0.3) ⎫	1.1[b]
Rivers, estuaries and coastal zones	1.7 (0.5–2.9) ⎭	
Biomass & biofuel burning	0.7 (0.2–1.0) ⎫	1.9[c]
Industry	0.7 (0.2–1.8) ⎬	
Atmospheric deposition	0.6 (0.3–0.9) ⎭	
Oceans	–[e]	1.0[d]
Total	6.7	7.8

Note: Values in brackets represent the uncertainty range.

[a] Direct emissions from agriculture

[b] Indirect emissions from agriculture, using revised emission factors[3,4]

[c] Includes N_2O from atmospheric N deposition, which is part agricultural in source

[d] Estimate of anthropogenic N_2O emission from oceans[5]

[e] Oceanic source denoted as 'Natural'[2]

tonnes of nitrogen (Tg N_2O-N) per year[1,11]. Though the big rise in emissions from human activities did not begin until the mid-19th century, it is estimated that the expansion of agriculture was already responsible for an extra one million tonnes of nitrous oxide-N per year by 1850, and that this expansion had actually replaced some of the natural sources of nitrous oxide on the land – reducing natural emissions by about 600,000 tonnes of nitrous oxide-N per year[12]. The natural sources of nitrous oxide are dominated by emissions from soils, especially from tropical soils, with the remainder coming from the oceans and some production in the atmosphere itself[1,13]. The estimates for these sources are quite uncertain though, as there are relatively few measurements for some sources (e.g. the oceans)[14] and untangling natural emissions from those caused by human activities can be very difficult.

Natural sources

Soils

Natural soils are the single largest source of nitrous oxide globally, currently responsible for around six million tonnes of nitrous oxide-N emission per year. Of this, natural soils in the tropics are estimated to add about four million tonnes[7,15], with around three million tonnes

---

Box 4.1

Ice core nitrous oxide

Ice cores are one of the most powerful tools we have in reconstructing past climates. As snow falls in the Arctic and Antarctic, it traps air between the flakes. As the snow builds up, the pressure increases and the snow gets compacted. Over hundreds of years the very compacted snow becomes ice and traps bubbles of air within it. As the ice sheets in Greenland and Antarctica are hundreds of thousands of years old, their ice contains a record of the air (and the nitrous oxide concentrations in it) going back for millennia. By drilling down through the ice we can go back in atmospheric time. Some of these ice cores are retrieved from two miles down in the ice sheet – they are under such immense pressure that care has to be taken that they do not explode once they are extracted.

By extracting the air from the ice at different depths the changes in nitrous oxide concentrations over millennia can be assessed. Because it takes a century or so for fallen snow to transform into ice (the time it takes for so much additional snow to build up above it and 'cap it off' from the atmosphere – at a depth between 60 and 100 m), the precision of measurements from ice cores is usually around 100 years. The most famous ice core is the Vostok ice core from East Antarctica[10] and this has been used to reconstruct the atmospheric concentrations of greenhouse gases for the last 420,000 years.

---

coming from wet forest soils and the rest arising from the soil of tropical dry savannas. Microbial nitrification or denitrification forms the basis of nitrous oxide production in both environments (Figure 4.1) – the moist, lower-oxygen environment of the wet forest soils often being conducive to denitrification[16].

Temperate soils are estimated to add a further two million tonnes of nitrous oxide-N to the atmosphere each year, with half coming via forest soils and the other half being emitted from the soils of temperate grasslands[15]. Again, the nitrous oxide arises via either nitrification or denitrification. Because the nitrification process relies on good availability of oxygen it is most important in well-drained and aerated soils, while the anaerobic conditions suitable for denitrification to occur become more prevalent in wetter or more compacted soils[16].

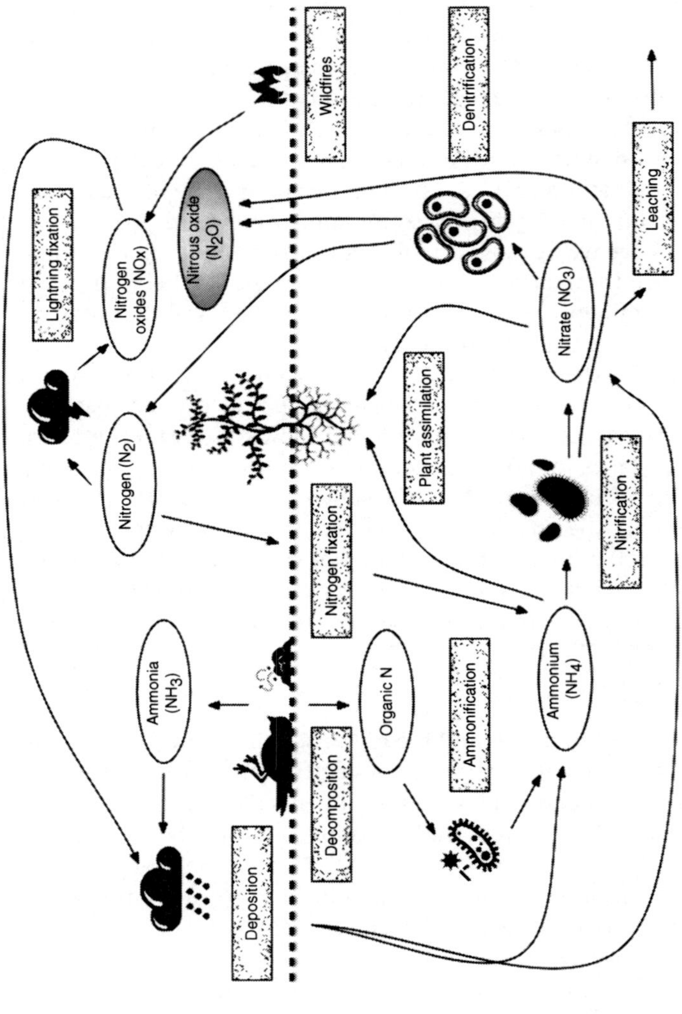

Figure 4.1   Nitrous oxide and the natural terrestrial nitrogen cycle

The dotted line represents the land surface with the boxes showing the key processes by which nitrogen is cycled and the ovals showing the changing form of nitrogen as it undergoes these processes. Nitrification and denitrification are the primary sources of nitrous oxide production.

Source: Dave Reay

Oceans

The current estimate of natural emissions from the oceans is around three million tonnes of nitrous oxide-N per year[15], but these estimates remain very uncertain because of the shortage of continuous measurements over such a vast area. Separating out the human-induced emissions is also a difficult challenge, with reactive nitrogen from human activities being introduced to the oceans both by rivers and by deposition from the atmosphere[17,18]. Whether a particular area of ocean acts as a source or a 'sink' of nitrous oxide depends on the level of nitrous oxide saturation in the water. Where it is fully saturated (100 per cent) it is deemed as being in equilibrium with the atmosphere, and so will be neither a net sink nor a net source. If the saturation rises above 100 per cent (called super-saturation) then the water becomes a source of nitrous oxide to the atmosphere; if it drops below 100 per cent it becomes a 'sink'. Samples of ocean water taken from around the world show that, as a whole, the oceans are slightly super-saturated (~104 per cent) and so a net source of nitrous oxide to the atmosphere[19].

As on the land, the primary pathways for nitrous oxide production in the oceans are via nitrifiers and denitrifiers[19], with nitrification responsible for about two-thirds of the production and denitrification accounting for the other third (Figure 4.2). As the pathway and rate of production is dependent on factors such as the form of reactive nitrogen, oxygen availability and temperature, the amount of nitrous oxide emitted can vary greatly from place to place and over time. In some coastal areas, where nitrogen-rich water is provided by the upwelling of deep ocean currents, the level of nitrous oxide saturation can reach several 1,000 per cent and so create a very strong source of nitrous oxide to the atmosphere. In others, where the supply of nitrogen is low or all the nitrous oxide is reduced to nitrogen gas – as can happen in completely anoxic waters – then the ocean area may become a sink rather than a source[20].

The ocean's nitrifiers and denitrifiers can be present both in the water and in the sediments below it. Because of their sensitivity to oxygen concentrations, however, nitrifiers tend to be more important in well-oxygenated waters and denitrifiers more important in the low-oxygen environments like sediments. For shallow coastal waters, nitrous oxide produced in sediments and then transferred up through the water to the atmosphere can be an important pathway for emissions. Out in the open ocean nitrous oxide may be produced within the water column and then emitted at the surface – windy conditions can greatly speed up this gas transfer and emission rate[21]. In deep waters nitrous oxide

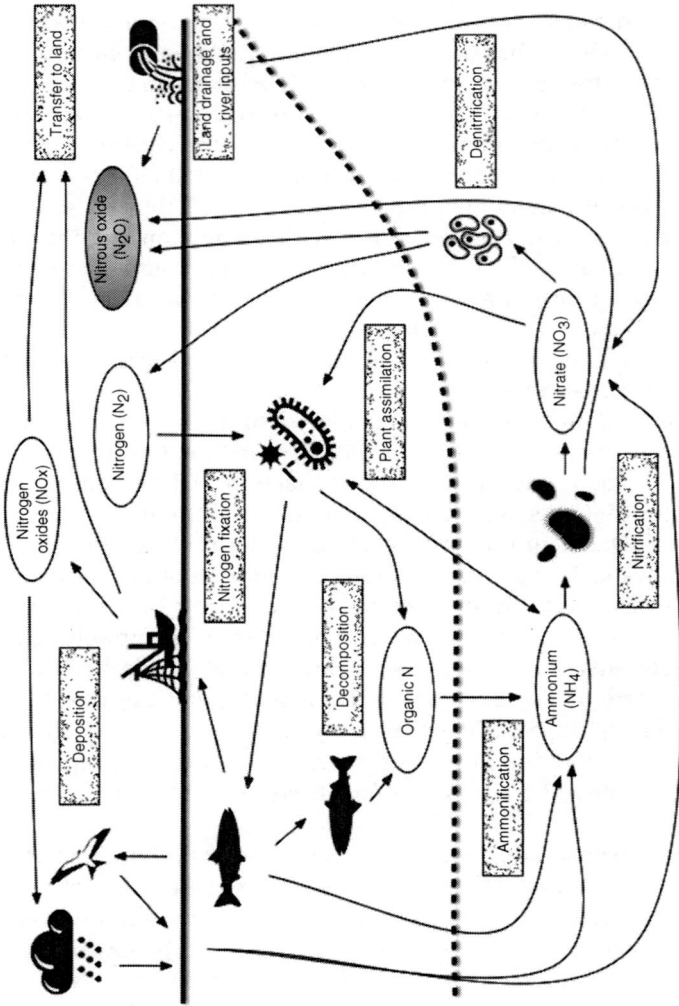

Figure 4.2   Nitrous oxide and the marine nitrogen cycle

The solid line represents the ocean surface and the dotted line the ocean floor. The boxes show the key processes by which nitrogen is cycled and the ovals show the changing form of nitrogen as it undergoes these processes. Nitrification and denitrification are the primary sources of nitrous oxide production.

Source: Dave Reay

may also be formed and then transported long distances by ocean currents before being released when the water eventually upwells to the surface.

Atmosphere

Although around 600,000 tonnes of nitrous oxide-N are thought to be produced each year in the atmosphere itself[7,15], the mechanisms responsible for this and the extent to which it can really be called 'natural' are debatable. The primary pathway appears to be through the release of aerosols (small droplets) of reactive nitrogen into the air which then undergo a series of chemical reactions with sunlight, water and other aerosols to produce nitrous oxide[22]. Nitrogen aerosols such as ammonium nitrate, nitrogen dioxide and ammonia are all thought to contribute to the overall formation of nitrous oxide in the atmosphere, but as these can come from both natural and human sources, the global atmospheric source of nitrous oxide is more a mixture of natural and human-induced causes.

Anthropogenic sources

Agriculture

Agriculture is by the far the biggest source of human-induced nitrous oxide emissions around the world[1]. Currently it is responsible for around four million tonnes of nitrous oxide-N each year, with these emissions coming not only from farmed soils (called 'direct emissions'), but also from the reactive nitrogen that is lost to the air and water as a result of agricultural activity and ends up causing additional nitrous oxide emissions (called 'indirect emissions').

Most direct emissions come from the addition of nitrogen-rich fertiliser, crop residues or manure to soils. The soil microbial processes of nitrification and denitrification then result in conversion of some of this added nitrogen to nitrous oxide gas, which is then released directly back into the atmosphere from the soil surface (Figure 4.3). The main indirect emissions from agriculture arise from nitrogen leaching and runoff, ammonia deposition, nitrogen fixation by legume crops and the management of farm products and waste[1]. Though less obvious than the direct emissions from soil, these indirect routes can be very important sources – often responsible for more than 20 per cent of total emissions – and so need to be included if the full impact of agricultural nitrous oxide emissions on climate is to be estimated.

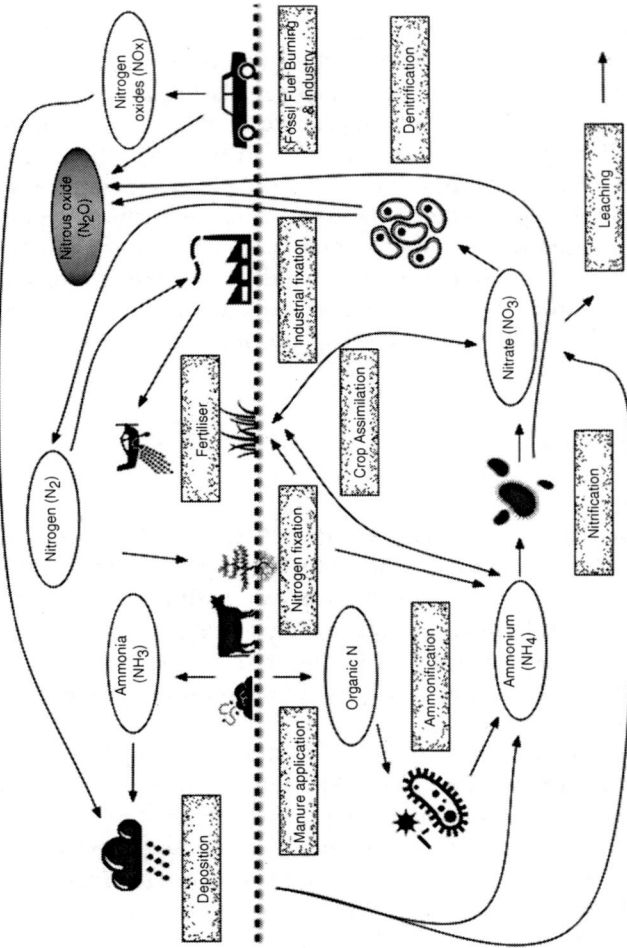

Figure 4.3 Anthropogenic nitrous oxide and the terrestrial nitrogen cycle

The dotted line represents the land surface with the boxes showing the key processes by which nitrogen is cycled and the ovals showing the changing form of nitrogen as it undergoes these processes. Much of the human-induced nitrous oxide arises from addition of nitrogen fertilisers to soils and the use of this added nitrogen in nitrification and denitrification. Some nitrous oxide is also released directly to the atmosphere from industry and fossil fuel burning.

Source: Dave Reay

Leaking nitrous oxide

The large fertiliser and manure additions made on farms around the world are aimed at increasing the growth of crops or pasture, but around half of this added nitrogen, and sometimes much more, is lost from the soils instead of being used by the plants for which it was intended.[23] These big losses occur because of a mismatch between the nitrogen needs of the plants and the timing, form and amount of nitrogen additions to the soils in which they are growing.

On many farms, artificial nitrogen fertilisers or manure is spread over the soil in two or three large applications each year. Once this wave of nitrogen fertiliser has landed on the soil its route into the crops is beset by a whole range of obstacles and blind alleys that can end in extra nitrous oxide emissions instead of extra plant growth.

Right away there are potential losses to the air through the process of volatilisation. Here, reactive nitrogen is lost from the fertiliser in the form of ammonia gas; this is then carried by the wind to be redeposited on land or water in another location[24]. Volatilisation is a particularly important problem in the use of ammonia-rich fertilisers, such as urea and manure, in warm and windy conditions. In such cases, losses to the air can amount to more than a third of the added nitrogen and the ammonia can travel several kilometres before it is redeposited[25]. When this ammonia does finally touch down, either on land or in water, it can go on to enhance nitrous oxide emissions through the processes of nitrification and denitrification. Globally, such indirect nitrous oxide emissions arising from atmospheric deposition of agricultural nitrogen are estimated to be 300,000 to 400,000 tonnes of nitrogen per year[26].

Back at the field, the reactive nitrogen that escapes volatilisation also faces the risk of being washed off or leached through the soil by heavy rain and ending up in groundwater, drainage streams and rivers[27]. The surface run-off pathway can be especially important where the fertiliser or manure is added to the surface of bare soil just before a period of very heavy rain. The leaching pathway is a major problem when the nitrogen fertiliser is in the form of nitrate – a very mobile form of nitrogen that is easily carried away as water drains through the soil[28]. On average, about one-third of the added fertiliser is lost due to such leaching[4].

Nitrous oxide emissions from nitrogen leaching and run-off take two main forms. The first is the release of any dissolved nitrous oxide to the atmosphere from drainage waters as soon as they leave the fields and enter open drainage ditches, streams and rivers[29]. This nitrous oxide is formed in the soil and in underground waters, and is then carried into drainage systems to eventually emerge in drainage ditches and streams. Because nitrous oxide is so soluble in water, it can quickly 'outgas' once it enters the open streams, with water carrying nitrous oxide at more than a hundred times the background concentration, losing all of its nitrous oxide load within just a few hundred metres[30]. Anything that mixes up the stream water makes this outgassing occur even faster, with waterfalls or weirs just downstream from the entry point of a nitrous oxide-loaded drain outlet being hot spots of emissions[31]. These emissions from dissolved nitrous oxide tend to occur relatively close to the fields and farms where the nitrous oxide was formed. However, such drainage waters can also contain large amounts of leached nitrate, and it is nitrate that can go on to be a source of nitrous oxide hundreds of miles downstream from the farm from which it was lost.

As the nitrate flows downstream and enters larger and larger rivers, it may pass over sediments containing denitrifiers that draw down some of the nitrate from the overlying water and emit nitrous oxide. As the waters approach the sea and the river speed slows, still more of the nitrate may be converted to nitrous oxide by denitrifiers in the sediments and attached to particles in the water[18]. Finally, once in estuaries and coastal seas, a combination of low-oxygen conditions and the inflowing nitrate from rivers can produce another burst of nitrous oxide production and emission[21]. Current estimates for indirect nitrous oxide emissions from leaching and run-off range from one million to two million tonnes N per year[26].

Agricultural nitrogen can therefore induce nitrous oxide emissions over a wide area and across long time-spans via losses to air and water. For the surviving fraction of the applied nitrogen that is retained in the farm soil, there remains the final hurdle of avoiding the many soil microbes that compete with the roots of the crops for gaining access to it. All soil microbes require some nitrogen, but some, especially the nitrifiers and denitrifiers, are able to intercept and transform a large proportion of the added fertiliser. As we have seen, the nitrifiers are able to convert ammonia to nitrate in the soil, so this can mean that yet more of the added nitrogen is lost via leaching. The denitrifiers can also use the nitrate and convert it into dinitrogen and nitrous oxide[28], diverting still more of the fertiliser nitrogen away from the crops and back into the atmosphere.

Though the added reactive nitrogen that is used by the agricultural crops and grasslands is now where it was intended to be and largely safe from causing still more nitrous oxide emissions, this safety is only temporary. On harvesting, the nitrogen locked up in any crop leaves, stems and roots that are left in the fields can be mineralised and again become available to the nitrifiers and denitrifiers[32]. Likewise for the legumes – such as bean and pea crops – the nitrogen the plants fix from the atmosphere as they grow provides a supply both for the plants themselves and for the soil around their roots[33]. The nitrifiers and denitrifiers in these soils can then take advantage of any such legume leakage and when, as often happens, such legume crops are ploughed back into the soil, a further burst of nitrous oxide production may occur.

For grasslands, much of the reactive nitrogen they now hold is taken in by the grazing livestock, digested and then redeposited on the land as manure and urine – again providing an opportunity for the nitrifiers and denitrifiers to make use of it[34]. For the nitrogen bound up in crops, grass and livestock that are then removed from the field, the subsequent uses of these agricultural products provide a multitude of additional opportunities for nitrous oxide emissions. Fodder crops and silage are fed to livestock with the resultant nitrogen-rich manure often collected and stored in large piles or silos. Large emissions of nitrous oxide can arise from the livestock sheds and feedlots, and from manure storage, with global emissions from such 'manure management' sources being around two million tonnes of nitrous oxide-N per year[15].

Similarly, much of the nitrogen in the crops and livestock products that are consumed by humans ends up in sewage treatment systems. These systems are hot spots of nitrous oxide emission – estimated to emit 200,000 tonnes of nitrous oxide-N per year globally – as nitrification and denitrification are deliberately used to remove nitrogen from the sewage during processing[35].

Bioenergy and aquaculture

In addition to crops for human and animal consumption, a large and growing amount produced by agriculture around the world is also used for biofuel production[36]. The main crops for this use are maize and sugar cane for the production of ethanol, and vegetable oils for biodiesel – in 2010 around 160 million tonnes of grain and vegetable oils were used for biofuels globally[37]. Such biofuels may appear attractive in terms of tackling climate change, as they can be substituted for petroleum and so reduce carbon dioxide emissions from fossil fuel burning[38]. Unfortunately, the cultivation of these so-called first-generation biofuels can lead to substantial nitrous oxide emissions and, in some cases, these emissions can offset some or all of the climate benefits of fossil fuel substitution[39]. There is a lack of good information on nitrous oxide from all biofuel production around the world and, with the growing demand for such fuels, emissions could rise in the coming decades[40].

Aquaculture has also become a major source of food globally and is increasingly substituting wild-caught fishing – it has been expanding at a rate of almost nine per cent per year since 1970[37]. Its impact on global nitrous oxide emissions remain poorly quantified; to date, there are only a handful of studies where emissions have been measured. However, many aquaculture systems involve intensive production and large amounts of reactive nitrogen inputs to aquatic systems via feeds and wastes. In such intensive conditions the emissions of nitrous oxide can be very high, with a current estimate of 120,000 tonnes of nitrous oxide-N per year emitted due to aquaculture globally[41].

Biomass burning

The final significant direct source of anthropogenic nitrous oxide is biomass burning[42], though its true magnitude remains highly uncertain (Box 4.2). It is currently estimated to produce around 500,000 tonnes of nitrous oxide-N each year[15], with earlier estimates being much higher than this due to the same errors in collection and analysis as those made for fossil fuel emissions[43]. Biomass burning occurs in many areas of the world and is often associated with land clearing for agriculture, disposal of straw and farm wastes, and domestic fuel burning (e.g. charcoal and wood-burning stoves). Nitrous oxide is produced during combustion of the nitrogen held within the biomass, with fires that burn at low temperatures or with a limited supply of oxygen tending to be the biggest emitters. Where the biomass is in the form of moist, nitrogen-rich material, such as manure, the nitrous oxide emissions can be especially high. Almost all biomass burning in the world is now human-induced, with a small fraction coming from natural fires caused by lightning strikes[44]. As well as the direct emissions of nitrous oxide from biomass burning, the pulse of reactive nitrogen (usually as ammonium) that is often released into the soil after burning can cause still more nitrous oxide to be produced, and additional emissions to the atmosphere from this indirect route[45].

Box 4.2

Estimating nitrous oxide emissions

Nitrous oxide emission factors are widely used to estimate emissions arising from a defined unit of a specific activity. Such estimates are used both for international reporting to the United Nations Framework Convention on Climate Change (UNFCCC) and for myriad national and sub-national reporting purposes, such as the European Union Emissions Trading Scheme (EU ETS). As with the other so-called 'Kyoto protocol greenhouse gases', the Intergovernmental Panel on Climate Change (IPCC) provides a methodology for national and sub-national estimation of nitrous oxide emissions, based on the sector from which the emissions arise[4]. Emissions are estimated using 'Tier 1, 2 or 3' methodologies, where Tier 1 relies on a universal emission factor combined with activity data, Tier 2 utilises a country-specific emission factor, and Tier 3 involves direct measurement or modelling approaches. For estimation of emissions from sources such as biomass burning or the agricultural sector, Tier 3 estimates are rarely available and default emission factors are often employed[46].

Energy and transport

Because of the rapid expansion of the energy sector in the 20th century and the fact that some nitrous oxide was emitted when fossil fuels were burned, this sector was once thought to be a major source of global nitrous oxide emissions. It was known that oil, gas and especially coal often contained small amounts of reactive nitrogen and that the process of burning these fuels could release some of this nitrogen in the form of nitrous oxide. More importantly, the high temperatures achieved during fossil fuel combustion could create nitrous oxide directly from the air by breaking down inert nitrogen gas and combining it with oxygen.

To measure just how much nitrous oxide was being emitted from the growing number of fossil-fuelled power stations, samples were taken from their chimneys and analysed. These so-called 'grab samples' appeared to show high concentrations of nitrous oxide coming from the burning of fossil fuels and so flagged up the energy sector as a significant global source[47]. Then, in the late 1980s, a mistake was spotted.

The way the grab samples had been collected and stored meant that extra nitrous oxide had actually been produced in the sample air. It turned out that, in most cases, the high temperatures at which the fuels were being burned (~900°C) meant that the bulk of any nitrous oxide produced was quickly destroyed well before it could escape to the atmosphere[48]. So, though the amounts of coal, oil and gas being burned were rising fast, the amount of nitrous oxide that was emitted from each tonne burned was actually around 100 times lower than what had been assumed. Some power stations that burn fuel at a relatively low temperature (~700°C) do still produce a significant amount of nitrous oxide[22], but on a global scale fossil fuel burning has dropped from being a major direct source of nitrous oxide to being a minor one, responsible for just three per cent of all the human-induced nitrous oxide emissions.

Transport and nitrogen pollution swapping

Aside from 'stationary sources' such as power stations, the other main source of nitrous oxide due to fossil fuel burning is transport[22]. As a very fast-growing sector, transport has the potential to be an increasingly large source of nitrous oxide emissions over the coming decades. There is also the extra complication caused by the introduction of catalytic convertors on many road vehicles, as these can actually increase how much nitrous oxide is emitted. As in the power stations, the nitrous oxide from road vehicles is produced both from oxidation of dinitrogen gas and from the trace amounts of nitrogen in the fuel that is burned, with most of this then destroyed by the high temperatures in the vehicle engines before it can escape via the exhaust pipe. The problem with this is that the forms of reactive nitrogen that do escape to the atmosphere – called nitrogen oxides or NOx – can themselves cause serious air pollution and human health problems. To control these problems in the face of big rises in vehicle use in the 1980s and 1990s, catalytic convertors were introduced in new cars. These catalytic convertors were very good at reducing how much NOx was emitted and so helped address the local air pollution problems, but they also meant that nitrous oxide emissions could be much greater – thereby swapping a local air pollution problem for a global climate change problem[49].

As the new vehicles fitted with catalytic convertors began to spread around the world in the early 1990s, more and more concerns were raised about the effect this might have on nitrous oxide emissions. Initial tests on traffic passing through tunnels in Sweden and Germany suggested that the new vehicles were indeed emitting much more than the old non-catalytic ones[50]. If these tests were correct, then, as all the old vehicle fleet was replaced with the new vehicles, nitrous oxide emissions from transport could grow to a level where this source would be responsible for a third of the global increase in atmospheric concentrations each year. In fact, subsequent measurements showed that the extra nitrous oxide emissions from vehicles with catalytic convertors tended to be much lower than it was first thought, with the total contribution of road vehicles to growth in nitrous oxide concentrations in the atmosphere being 1 to 4 per cent rather than 30 per cent[51]. For aircraft too, the few measurements of nitrous oxide emissions that exist suggest that while some planes can emit substantial amounts individually, taken as a whole air travel remains a minor global source[52]. Overall, direct nitrous oxide emissions from transport are estimated to be 100,000 to 250,000 tonnes of nitrous oxide-N per year.

Industry

Industrial sources are thought to emit around 1.3 million tonnes of nitrous oxide-N into the atmosphere each year[15]. Of this, nitric acid production is one of the most important. Nitric acid is a key ingredient in nitrogen-based fertilisers and the manufacture of explosives, adipic acid (for making nylon) and various metals[22]. Its method of production, called the Ostwald process, involves the oxidation of ammonia using a platinum catalyst and high temperatures[53]. Nitrous oxide arises from this process during the catalytic oxidation step, and it is estimated that around five grams are produced in this way for every kilogram of nitric acid that is obtained. The higher the temperature used, the greater the amount of ammonia that is transformed into nitrous oxide, and in early production plants much of this would be emitted into the atmosphere via their chimneys. In the last two decades much progress has been made in reducing the nitrous oxide emissions from this process by using more efficient conditions for ammonia oxidation and installing waste gas 'scrubbers' that contain iron-based catalysts and decompose up to 99 per cent of the nitrous oxide before it escapes[54].

The other major source of industrial nitrous oxide emissions is that of nylon manufacture and specifically the production of adipic acid, which is a fine powder that is used in manufacturing nylon and for making dyes and insecticides. Nitrous oxide is produced during the oxidation of a ketone–alcohol mixture with nitric acid, and it is estimated that for each kilogram of adipic acid made, around 30 grams of nitrous oxide is also produced[22]. Again, in earlier production plants much of the nitrous oxide would then be released into the atmosphere[55]. However, since the mid-1990s big improvements have been made to ensure the removal of nitrous oxide before emission into the atmosphere. This involves its decomposition using either catalysts or high temperatures, with most plants having now installed these technologies and achieving cuts of over 90 per cent in emissions from this source[56].

Nitrogen trifluoride

Though a different form of reactive nitrogen, nitrogen trifluoride, also emitted by industrial sources, is a powerful greenhouse gas that is becoming increasingly important due to its very high global warming potential (GWP). Nitrogen trifluoride is used in processes such as the manufacture of liquid crystal displays and some solar panels; it has a GWP of 17,200 and, in 2013, was added to list of greenhouse gases that each nation must report on each year in its annual inventory of emissions. The concentration of nitrogen trifluoride in the atmosphere remains very low, and its climate impact is far lower than that of nitrous oxide. Nevertheless, with such a high GWP value and fast growth in its usage, this rather exotic nitrogen-rich greenhouse gas is one that may become more and more important in the coming years.

Nitrogen deposition and indirect nitrous oxide

Just as airborne reactive nitrogen emitted from agriculture (e.g. ammonia) can be redeposited on land and water to promote nitrous oxide emissions, emissions from fossil fuel combustion and industry (e.g. nitrogen oxides) can also be major indirect sources of nitrous oxide. Global emissions of nitrogen oxides from fossil fuel combustion now exceed 25 million tonnes of nitrogen each year, with around half the deposits being on the land and the rest ending up falling on coastal seas and the open ocean[57,58]. Assuming that between one and two per cent of this rain of reactive nitrogen is then converted to nitrous oxide (Box 4.2), fossil fuel burning may indirectly add a further 500,000 tonnes of nitrous oxide to the atmosphere each year. Such nitrogen-based greenhouse gas emissions would therefore add the equivalent of an extra 230 million tonnes of carbon dioxide to the already-large climate impact of global fossil fuel burning.

Likewise, the reactive nitrogen emissions arising from biomass burning serve to magnify its climate-forcing impacts. Current nitrogen oxide emissions from this source are estimated to be around six million tonnes of nitrogen per year[58] – resulting in enhanced nitrous oxide emissions from land and water of up to 300,000 tonnes of nitrogen a year.

Nitrous oxide sinks

Nitrous oxide sinks are areas where the net flux results in a removal of nitrous oxide from the atmosphere, rather than an addition to it[13]. The majority of atmospheric nitrous oxide is destroyed in the stratosphere (the upper atmosphere) by reaction with light and excited oxygen atoms, with this sink accounting for around 13 million tonnes of nitrous oxide-N each year[15]. At the Earth's surface, the land taken as a whole is a net source, but within this there are some areas that appear to act as net sinks over limited spatial scales and time spans[7]. The mechanism behind this soil uptake of nitrous oxide from the atmosphere remains unclear, and work is continuing on trying to understand and quantify it properly. Because the individual sink areas can be very small (<1 m) and short-lived (<1 h), standard measurement techniques often struggle to disentangle them from the wider background of soil nitrous oxide emissions. One suggestion is that it is the soil denitrifiers and the conditions they experience that determine whether the switch from source to sink occurs. Denitrifiers are able to reduce nitrous oxide to nitrogen gas; hence, where nitrous oxide is available and conditions for the denitrifiers to use it are optimal, they may be able to deplete the nitrous oxide in the surface layers of the soil and therefore cause more to diffuse in from the air above[3]. Certainly, such denitrification can play a key role in limiting how much of the nitrous oxide that is produced in soils is actually emitted to the atmosphere, with the near-surface denitrifiers intercepting nitrous oxide as it diffuses up from deeper layers. So, even where the net effect is still a nitrous oxide source, the reduction of nitrous oxide to nitrogen gas by soil denitrifiers represents a crucial buffer for atmospheric emissions – without it emissions would be up to 10 times what they actually are.

References

1. Bouwman, L. et al. Drawing down N_2O to protect climate and the ozone layer. A UNEP Synthesis Report. (2013).
2. Solomon, S. *Climate change 2007 – the physical science basis: Working Group I contribution to the fourth assessment report of the IPCC.* Vol. 4. (Cambridge University Press, 2007).
3. Kroeze, C., Bouwman, L. & Slomp, C. P. Sinks for nitrous oxide at the earth's surface. In *Greenhouse Gas Sinks*, edited by D. Reay, C. N. Hewitt, K. Smith & J. Grace, 227–242 (CABI, Wallingford, UK, 2007).
4. Eggleston, S., Buendia, L., Miwa, K., Ngara, T. & Tanabe, K. IPCC guidelines for national greenhouse gas inventories. *Institute for Global Environmental Strategies, Hayama, Japan* (2006).

5. Duce, R. A. et al. Impacts of atmospheric anthropogenic nitrogen on the open ocean. *Science* **320**, 893–897, doi:10.1126/science.1150369 (2008).

6. Alley, R. B. *The two-mile time machine: ice cores, abrupt climate change, and our future*. (Princeton University Press, 2000).

7. Stocker, T. *Climate change 2013: the physical science basis: Working Group I contribution to the fifth assessment report of the Intergovernmental Panel on Climate Change*. (Cambridge University Press, 2014).

8. Wolff, E. & Spahni, R. Methane and nitrous oxide in the ice core record. *Philosophical Transactions of the Royal Society A: Mathematical, Physical and Engineering Sciences* **365**, 1775–1792 (2007).

9. Sowers, T., Alley, R. B. & Jubenville, J. Ice core records of atmospheric N_2O covering the last 106,000 years. *Science* **301**, 945–948 (2003).

10. Petit, J.-R. et al. Climate and atmospheric history of the past 420,000 years from the Vostok ice core, Antarctica. *Nature* **399**, 429–436 (1999).

11. Davidson, E. A. The contribution of manure and fertilizer nitrogen to atmospheric nitrous oxide since 1860. *Nature Geoscience* **2**, 659–662, doi:10.1038/ngeo608 (2009).

12. Syakila, A. & Kroeze, C. The global nitrous oxide budget revisited. *Greenhouse Gas Measurement and Management* **1**, 17–26 (2011).

13. Reay, D. S., Hewitt, C. N. & Smith, K. A. Nitrous oxide: importance, sources and sinks. In *Greenhouse Gas Sinks*, edited by D. Reay, C. N. Hewitt, K. Smith & J. Grace, 201–206, doi:10.1079/9781845931896.0201 (CABI, Wallingford, UK, 2007).

14. Duce, R. et al. Impacts of atmospheric anthropogenic nitrogen on the open ocean. *Science* **320**, 893–897 (2008).

15. Smith, K., Crutzen, P., Mosier, A. & Winiwarter, W. The global nitrous oxide budget: a reassessment. In *Nitrous Oxide and Climate Change*, edited by K. Smith, 63–84 (Earthscan, London, 2010).

16. Davidson, E. A. Soil water content and the ratio of nitrous oxide to nitric oxide emitted from soil. In *Biogeochemistry of Global Change*, edited by R. S. Oremland, 369–386 (Chapman & Hall, 1993).

17. Kroeze, C., Seitzinger, S. P. & Domingues, R. Future trends in worldwide river nitrogen transport and related nitrous oxide emissions: a scenario analysis. *The Scientific World Journal* **1 Suppl. 2**, 328–335, doi:10.1100/tsw.2001.279 (2001).

18. Kroeze, C. & Seitzinger, S. P. Nitrogen inputs to rivers, estuaries and continental shelves and related nitrous oxide emissions in 1990 and 2050: a global model. *Nutrient Cycling in Agroecosystems* **52**, 195–212 (1998).

19. Bange, H. W., Freing, A., Kock, A. & Löscher, C. Marine pathways to nitrous oxide. *Nitrous Oxide and Climate Change*, 36–54 (Earthscan, New York, 2010).

20. Codispoti, L. & Christensen, J. Nitrification, denitrification and nitrous oxide cycling in the eastern tropical South Pacific Ocean. *Marine Chemistry* **16**, 277–300 (1985).

21. Nevison, C. D., Weiss, R. F. & Erickson, D. J. Global oceanic emissions of nitrous oxide. *Journal of Geophysical Research: Oceans (1978–2012)* **100**, 15809–15820 (1995).

22. Wiesen, P. Abiotic nitrous oxide sources: chemical industry and mobile and stationary combustion systems. In *Nitrous Oxide and Climate Change*, edited by K. Smith, 190–209 (Earthscan, London, 2010).

23. Erisman, J. W. et al. The European nitrogen problem in a global perspective. In *The European Nitrogen Assessment: Sources, Effects and Policy Perspectives*, edited by M. A. Sutton et al., 9–31 (Cambridge University Press, UK, 2011).

24. Jarvis, S. & Pain, B. Ammonia volatilization from agricultural land. *Proceedings – Fertiliser Society* 2, No. 298, 35 pp. (CABI, Wallingford, UK, 1990).

25. Butterbach-Bahl, K. et al. Nitrogen processes in terrestrial ecosystems. In *The European Nitrogen Assessment: Sources, Effects and Policy Perspectives*, edited by M. A. Sutton et al., 6, 99–125 (Cambridge University Press, UK, 2011).

26. Mosier, A. et al. Closing the global N(2)O budget: nitrous oxide emissions through the agricultural nitrogen cycle – OECD/IPCC/IEA phase II development of IPCC guidelines for national greenhouse gas inventory methodology. *Nutrient Cycling in Agroecosystems* 52, 225–248, doi:10.1023/a:1009740530221 (1998).

27. Galloway, J. N. et al. The nitrogen cascade. *Bioscience* 53, 341–356 (2003).

28. Smith, S., Schepers, J. & Porter, L. Assessing and managing agricultural nitrogen losses to the environment. *Advances in Soil Science*, edited by B. A. Stewart, 14, 1–43 (Springer, 1990).

29. Reay, D. S., Smith, K. A. & Edwards, A. C. Nitrous oxide emission from agricultural drainage waters. *Global Change Biology* 9, 195–203, doi:10.1046/j.1365-2486.2003.00584.x (2003).

30. Reay, D. S., Edwards, A. C. & Smith, K. A. Importance of indirect nitrous oxide emissions at the field, farm and catchment scale. *Agriculture Ecosystems & Environment* 133, 163–169, doi:10.1016/j.agee.2009.04.019 (2009).

31. Reay, D., Edwards, A. & Smith, K. Determinants of nitrous oxide emission from agricultural drainage waters. *Water, Air, & Soil Pollution: Focus* 4, 107–115 (2004).

32. Velthof, G. L., Kuikman, P. J. & Oenema, O. Nitrous oxide emission from soils amended with crop residues. *Nutrient Cycling in Agroecosystems* 62, 249–261 (2002).

33. Rochette, P. & Janzen, H. H. Towards a revised coefficient for estimating N_2O emissions from legumes. *Nutrient Cycling in Agroecosystems* 73, 171–179 (2005).

34. Oenema, O. et al. Trends in global nitrous oxide emissions from animal production systems. *Nutrient Cycling in Agroecosystems* 72, 51–65 (2005).

35. Kampschreur, M. J., Temmink, H., Kleerebezem, R., Jetten, M. S. & van Loosdrecht, M. Nitrous oxide emission during wastewater treatment. *Water Research* 43, 4093–4103 (2009).

36. Ajanovic, A. Biofuels versus food production: does biofuels production increase food prices? *Energy* 36, 2070–2076 (2011).

37. FAOSTAT FAO Statistical database. *Food and Agriculture Organization of the United Nations*, http://faostat.fao.org (2013).

38. Bessou, C., Ferchaud, F., Gabrielle, B. & Mary, B. Biofuels, greenhouse gases and climate change. In *Sustainable Agriculture Volume 2*, edited by E. Lichtfouse et al., 365–468 (Springer, 2011).

39. Crutzen, P. J., Mosier, A. R., Smith, K. A. & Winiwarter, W. N_2O release from agro-biofuel production negates global warming reduction by replacing fossil fuels. *Atmospheric Chemistry and Physics* 8, 389–395 (2008).

40. Reay, D. S. Not so sweet after all? *Nature Climate Change* 1, 174–174 (2011).

41. Williams, J. & Crutzen, P. Nitrous oxide from aquaculture. *Nature Geoscience* 3, 143–143 (2010).
42. Crutzen, P. J. & Andreae, M. O. Biomass burning in the tropics: impact on atmospheric chemistry and biogeochemical cycles. *Science* 250, 1669–1678 (1990).
43. Cofer III, W., Levine, J., Winstead, E. & Stocks, B. New estimates of nitrous oxide emissions from biomass burning. *Nature* 349, 689–691 (1991).
44. Davis, M., Crowcier, C. & Richardson Jr, J. Importance of biomass burning in the atmospheric budgets of nitrogen-containing gases. *Nature* 346, 552–554 (1990).
45. Anderson, I. C., Levine, J. S., Poth, M. A. & Riggan, P. J. Enhanced biogenic emissions of nitric oxide and nitrous oxide following surface biomass burning. *Journal of Geophysical Research: Atmospheres (1984–2012)* 93, 3893–3898 (1988).
46. Reay, D. S. et al. Global agriculture and nitrous oxide emissions. *Nature Climate Change* 2, 410–416, doi:10.1038/nclimate1458 (2012).
47. Pierotti, D. & Rasmussen, R. Combustion as a source of nitrous oxide in the atmosphere. *Geophysical Research Letters* 3, 265–267 (1976).
48. Hayhurst, A. & Lawrence, A. Emissions of nitrous oxide from combustion sources. *Progress in Energy and Combustion Science* 18, 529–552 (1992).
49. Dasch, J. M. Nitrous oxide emissions from vehicles. *Journal of the Air & Waste Management Association* 42, 63–67 (1992).
50. Berges, M., Hofmann, R., Scharffe, D. & Crutzen, P. Nitrous oxide emissions from motor vehicles in tunnels and their global extrapolation. *Journal of Geophysical Research: Atmospheres (1984–2012)* 98, 18527–18531 (1993).
51. Becker, K. et al. Nitrous oxide (N_2O) emissions from vehicles. *Environmental Science & Technology* 33, 4134–4139 (1999).
52. Wiesen, P., Kleffmann, J., Kurtenbach, R. & Becker, K. Nitrous oxide and methane emissions from aero engines. *Geophysical Research Letters* 21, 2027–2030 (1994).
53. Hunt, L. The ammonia oxidation process for nitric acid manufacture. *Platinum Metals Review* 2, 129–134 (1958).
54. Pérez-Ramírez, J., Kapteijn, F., Schöffel, K. & Moulijn, J. Formation and control of N_2O in nitric acid production: where do we stand today? *Applied Catalysis B: Environmental* 44, 117–151 (2003).
55. Thiemens, M. H. & Trogler, W. C. Nylon production: an unknown source of atmospheric nitrous oxide. *Science* 251, 932–934 (1991).
56. Shimizu, A., Tanaka, K. & Fujimori, M. Abatement technologies for N_2O emissions in the adipic acid industry. *Chemosphere – Global Change Science* 2, 425–434 (2000).
57. Reay, D. S., Dentener, F., Smith, P., Grace, J. & Feely, R. A. Global nitrogen deposition and carbon sinks. *Nature Geoscience* 1, 430–437, doi:10.1038/ngeo230 (2008).
58. Jaeglé, L., Steinberger, L., Martin, R. V. & Chance, K. Global partitioning of NOx sources using satellite observations: relative roles of fossil fuel combustion, biomass burning and soil emissions. *Faraday Discussions* 130, 407–423 (2005).

5
Airborne Nitrogen and Climate Change

Nitrogen oxide sources

Nitric oxide (NO) and nitrogen dioxide (NO_2) are both gases and are commonly grouped together as NOx (pronounced 'nox'). Like nitrous oxide, the NOx gases are oxidised forms of nitrogen, with fossil fuel burning, biomass burning and cultivated soils being their largest anthropogenic sources[1,2] (Figure 5.1). Unlike nitrous oxide, these gases are highly reactive, with short atmospheric lifetimes and the ability to cause severe illness and even death in humans. Early fossil fuel-driven emissions of these NOx gases were dominated by the release of the trace amounts of reactive nitrogen contained in coal as it was burned[3]. With the advent of higher temperature boilers and the rapid spread of the internal combustion engine, more and more NOx was produced by the direct reaction of dinitrogen gas with oxygen – the high temperatures in power station furnaces and vehicle engines breaking apart the twin atoms of nitrogen gas and combining them with oxygen. The NOx gases that flow from exhausts and chimneys are usually in the highly reactive form of nitric oxide. Being so very reactive, this gas quickly combines with more oxygen and is converted into the brownish, acrid-smelling gas called nitrogen dioxide. The amounts released from burning fossil fuels have rocketed over the past century, especially since the Second World War. As the sulphur-enriched smogs of the 1950s began to clear, the injections of NOx into the atmosphere became more intense[4]. Between the 1960s and 1980s these human-made emissions more than doubled to 25 million tonnes of nitrogen per year and today make up over half of all the NOx produced across the planet[5].

Much of the world's soil-derived NOx now also has the fertile fingerprints of humankind upon it[6]. The busy microbial push-me pull-you of

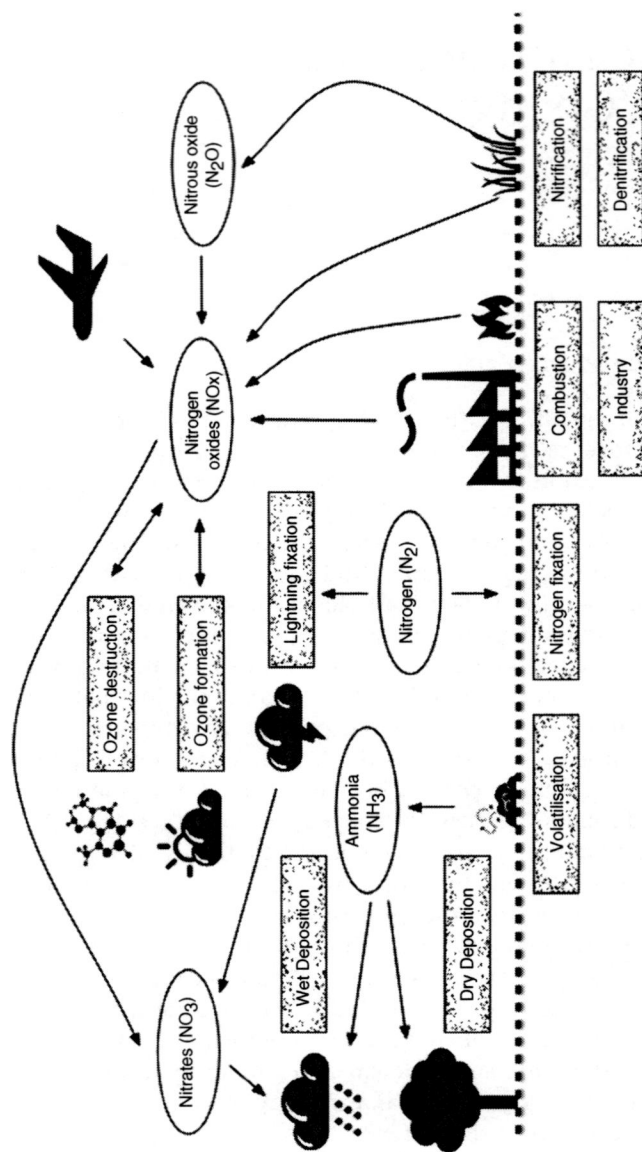

Figure 5.1   The atmospheric nitrogen cycle

The dotted line represents the land surface with the boxes showing the key processes by which nitrogen is cycled and the ovals showing the changing form of nitrogen as it undergoes these processes. The large pool of dinitrogen gas (N_2) in the atmosphere is constantly replenished through the microbial processes of denitrification and 'anammox'.

Source: Dave Reay

the nitrifiers and denitrifiers leads to NOx being produced and a small proportion of this key stepping stone in each process leaks away into the atmosphere. With more ammonium for the nitrifiers and more nitrate for the denitrifiers, fertilised-soil NOx emissions have kept on climbing, and now make up around half of all soil emissions globally[7].

Other important global sources of NOx include the air-splitting chemical super-highways of lightning, forest and grassland fires, and uncultivated soils. These three occurrences usually come under the umbrella of 'natural sources', yet much of the biomass burning that now goes on is far from natural, the fires that unlock and oxidise the nitrogen contained in plant tissues more often starting with a match strike than a thunderbolt[8].

NOx and human health

In regions where a great deal of NOx emission occurs in areas of high population density – as in many traffic-jammed cities – the air at ground level can become very nitrogen-rich and dangerous. Nitrogen dioxide is toxic to humans[9], and if its concentration in the air tops 150 parts per million it will begin to kill those who breathe it. Death can be rapid, with the lungs swelling and filling with fluid, or may take several weeks as the fine branches of the lungs – the bronchioles – become so inflamed that air supply is shut off and the victim slowly suffocates. Even at much lower levels, chronic exposure to nitrogen dioxide can cause serious damage to the lungs, and its presence has been implicated in cases of coronary heart disease and sudden infant death syndrome[10]. The direct effects of more NOx can therefore be severe, and efforts to reduce its emissions have intensified in the past few decades. But, for the health department of almost every major city in the world more NOx in the air heralds something even more dangerous: ozone. Ozone, that precious high-altitude gas that shields our planet and everything that lives on it from the most damaging rays of the sun, is also formed closer to earth by an excitable combination of NOx gases, other air pollutants and the crucial active ingredient of sunlight (collectively called 'ozone precursors')[11]. It is this potent mixture that helps give rise to the modern-day summertime smogs now endured by so many large urban areas (Box 5.1). Often called 'photochemical smog', ozone is its most infamous constituent and can cause severe internal burning and respiratory damage if breathed in[12,13]. As the amounts of ozone in the air begin to increase, so the damage to the lungs of those breathing it becomes more severe: at just one part in every million the passages of the lungs become irritated and

---

Box 5.1

The evolution of smog

For many, the word smog – a choking mix of smoke and fog – still brings to mind visions of the infamous 'pea soupers' of post-war London and the 'Auld Reekie' tag of Daniel Rutherford's Enlightenment Edinburgh. Archive footage of London's last pea souper in the early 1950s shows red double-decker buses struggling past street lights failing to penetrate the midday gloom, and the ghostly forms of pedestrians clutching handkerchiefs to their faces in an effort to filter the throat-burning air. These early and mid-20th century smogs had their root in the huge amounts of coal being burnt in the cities and in the sulphur it contained. When burnt, sulphur is released as sulphur dioxide, a gas that turns into sulphuric acid when oxidised and combined with the water in rain, in mist or in the moist linings of human lungs. The huge amounts of smoke particles and sulphur dioxide pouring from millions of city chimneys meant that on calm days just breathing the air could be life-threatening[15]. The thousands of smog-related deaths during the big London pea souper of 1952 prompted a clamp-down on coal burning in the city, and over the following decades a switch to gas and electricity helped to lift the threat of coal-smog from many of the world's biggest cities. But in their busy streets a new smog threat was fast emerging – instead of thousands of coal fires, the growing problem now was thousands of vehicle exhausts and the nitrogen oxides they emitted.

---

inflamed; at two parts per million even short periods of exposure can cause severe lung damage and death.

A further potential human health impact of NOx comes via its conversion into small particles of airborne nitrate through reaction with nitric acid. The risks from such particulate matter (PM) largely depend on the size of the particles, with the smallest being the most dangerous. For PM containing nitrate, little evidence exists on whether these particles pose a significant threat to human health, but it is known that they can contribute to the mix of fine particles in the air we breathe[14].

The cocktail of chemicals in the air over a big city is a continually changing sea of reactions as new gases and particles are released into it by the exhausts and chimneys below, and others are removed from it by wind, rain and gravity. There are myriad pathways for NOx gases to follow and some – like ozone formation – will open up only with the helping hand of sunlight. Things start to get really dangerous for human health where heavy NOx pollution coincides with long periods of calm, hot and sunny weather. As the concentrations of NOx and other pollutants increase, their molecules are more likely to come into contact with one another, and react. With strong sunlight and high temperatures to push along these reactions, more and more ozone is produced[16]. This photochemical smog – the late-20th-century variation of pea soupers – is now a global problem, and some cities, like Los Angeles and Mexico City, have its key ingredients in spades. Mexico City is hemmed in on all sides by hills, providing an ideal low-altitude mixing bowl for the pollutants that rise from below and the sunlight that streams in from above[17]. To produce ozone, the NOx gases react in sunlight with another group of pollutants called 'volatile organic compounds' (VOCs) (Box 5.2)[18]. This broad church of quick-to-vaporise substances can come from everything from paint stripper and carpets to glue sticks and petroleum, with plants and trees being an especially important source outside cities[19].

To try and limit the damage of enhanced urban ozone creation, monitoring and early warning systems are now in place in most cities of the world, with schools being closed and factories shut down if ozone levels creep too high[20]. More proactive regulation and curbs on vehicular NOx emissions have also helped to limit the frequency of killer smogs[21], while the improvement of catalytic convertor technology in cars and

---

Box 5.2

NOx as an urban ozone precursor

As the sun rises and its rays begin to hit the miasma of gases and particles hanging over the city, the initial reaction that ends with more ozone is actually to destroy some of it. In the presence of water vapour, the sunlight breaks apart those ozone molecules already present to produce a super-reactive product called 'hydroxyl radicals' (OH radicals). Some of these are mopped up by the NOx to make nitric acid, but others react with the VOCs to make yet another set of super-reactive products called 'peroxy radicals'. The peroxy radicals can in turn attack any nitric oxide and convert it into nitrogen dioxide. If the sun is shining and the temperature high, then this nitrogen dioxide is quickly split apart to produce a mix of nitric oxide and single oxygen atoms (O_1), which is the final key ingredient. These excited oxygen atoms (O_1) quickly combine with any free oxygen (O_2) available to make ozone (O_3)[18].

trucks over the last few decades has served to slash NOx emissions by converting the NOx gases back to nitrogen gas and oxygen before they leave the tailpipe.

In order to address bigger point sources of NOx, such as power stations and the nitric acid industry, waste gas scrubbing using high temperatures and/or catalysts is used to decompose the NOx before it can escape. Since the 1990s, tighter regulation of NOx emissions from these large point sources, in particular the Gothenburg Protocol, has led to significant decreases in most parts of the world[22,23]. These power plants also tend to be located further away from population centres so as to reduce their direct contribution to urban air pollution. They use tall chimneys to carry any NOx that does escape high into the air, to be dispersed over a wide area downwind, with the NOx gases then able to travel hundreds or even thousands of kilometres. At the point of emission the NOx is mostly in the form of nitric oxide, which actually mops up any ozone in the air, thereby ensuring that ozone concentrations close to the source are low. Further downwind, more and more of the nitric oxide is converted to nitrogen dioxide and ozone production begins to outweigh destruction. This spreading ripple of downwind ozone creation means that the human health impacts of NOx emissions from a power station can occur in villages, towns and cities in other countries or even in other continents[24].

Beating even such tall-chimneyed power stations for globe-trotting NOx impacts are emissions from aviation and international shipping[25]. NOx emissions from aircraft are particularly important, as the high-altitude conditions at which they occur (in the upper troposphere and lower stratosphere) are especially good for ozone formation compared to the conditions closer to the surface[26]. Indeed the large contribution of aviation NOx to ozone formation is a key reason why air travel has a significant impact on global climate[27]. International and regional limits on NOx emissions from shipping and air travel already exist, but the long lifespan of ships and planes means that many of the older and dirtier engines will take many years before they are replaced by a new fleet with lower NOx emissions.

Ammonia in the atmosphere

Producing ammonia (NH_3) from fossil fuel burning was at the heart of early efforts to find a plentiful supply of reactive nitrogen for food and explosives (see Chapter 2). Back then the aim was to make use of the trace amounts of reactive nitrogen contained in coal by releasing them through combustion in low-oxygen conditions, and then capturing the ammonia-rich gases by dissolving them in water. Fritz Haber's 1908 invention of ammonia synthesis has long since put paid to the need for such deliberate ammonia production from coal, but a side effect of burgeoning agricultural expansion since the industrial revolution has been a huge rise in global ammonia emissions[5,28]. Their main anthropogenic sources are fertiliser use and livestock farming[29], with emissions therefore predominantly occurring in the countryside. Some ammonia is still also emitted from fossil fuel burning and industry[30], but these sources are now dwarfed by global agriculture.

Ammonia is a strong-smelling gas that is quick to react with other atmospheric pollutants. It plays an important role in global climate and human health – the latter being primarily through its role in the production of small airborne particles, such as ammonium nitrate. As previously discussed for NOx, such PM in the air represents a major air pollution-derived human health problem globally, with particles less than 2.5 microns in size (called $PM_{2.5}$) being one of the most dangerous forms, as they can pass deep into the human respiratory system[31]. Around 20 per cent of all such 'secondary PM' is now derived from ammonia, with ammonia emissions and the particles they induce set to increase further in the coming decades. Again though, the lack of information on the human health impacts of specific PM components makes the precise effects of ammonia uncertain[14].

Climate change interactions

The impact of future climate change on concentrations of airborne nitrogen (NOx and ammonia) may take numerous forms, including the potential for enhanced emissions from wildfires, soils and manure, and a slight increase in lightning-induced NOx production[32]. Summertime increases in power station NOx emissions due to a greater electricity demand for air-conditioning have also been suggested; any such impacts that could exacerbate the problem of nitrogen-induced ozone formation are of particular concern.

As tropospheric ozone formation can increase with higher temperatures, lighter winds and more sunshine, the weather is a major factor in determining whether or not its levels become dangerous to humans[33]. Projected climate change is expected to increase both average and peak temperatures around the world during the 21st century, making higher rates of ozone creation in some areas during the 21st century very likely[34]. Cities and the people living in them are at particular risk from this climate-driven ozone effect due to their ready supply of NOx and other ozone precursors, their reduced wind speeds and the higher temperatures they experience compared to the surrounding countryside – the so-called 'urban heat island' (UHI) effect (Box 5.3).

Box 5.3

Urban heat islands

Because buildings and roads tend to be darker and less reflective than fields and forests, they absorb more of the sun's energy and emit more heat. The reflectiveness of a surface is termed its 'albedo', and dark surfaces such as roads and tiled roofs frequently have albedos as low as 0.1 – which means that they reflect just 10 per cent of the solar energy hitting them. This urban albedo effect, combined with waste heat from buildings and poor air mixing, can push the temperatures in cities and large towns much higher (often 5°C higher) than those in the surrounding countryside[35]. These urban heat islands (UHIs) of elevated temperature can therefore act as warm oases for ozone formation even when temperatures outside the city remain too low for rapid ozone production to become a problem. Alongside the elevated temperatures in the city, the tall and closely packed buildings help to lower wind speeds and stabilise the air, thereby allowing pollutants like NOx and the VOCs to build up more easily. Though tackling urban ozone by controlling NOx emissions, especially from vehicles, is a prime target for cities already badly affected by summertime high-ozone pollution events, there is also the potential to limit ozone formation by reducing the UHI effect itself[36]. The focus of this temperature-reduction approach is often on raising the albedo of the city surface to make it more reflective and so reduce overall temperatures. Urban temperatures can be greatly reduced by changing dark, low-albedo surfaces to more reflective materials, such as white-painted roofs and pavements. This has win-win benefits for the city's inhabitants, as it also reduces the energy demand for air-conditioning in buildings.

Using urban cooling strategies such as 'cool roofs' and 'cool pavements' can therefore represent an effective form of climate change adaptation in cities at risk from the deadly combination of summertime heat waves and ozone peaks. However, an alternative strategy for urban cooling that employs greater use of trees and vegetation may in fact increase the ozone problem even where it succeeds in reducing the UHI effect. Here, the shade offered by trees and the cooling effect of vegetation caused by its uptake and evaporation of water (transpiration) can be used to cool urban environments. More vegetation may increase the cooling effect, but is also a key source of the very VOCs that then interact with NOx and sunlight to create ozone. Whether more urban vegetation actually increases the ozone problem depends on the types of plant used – some, such as oak trees, are high VOC emitters – and the degree to which ozone formation is limited by VOC supply[37].

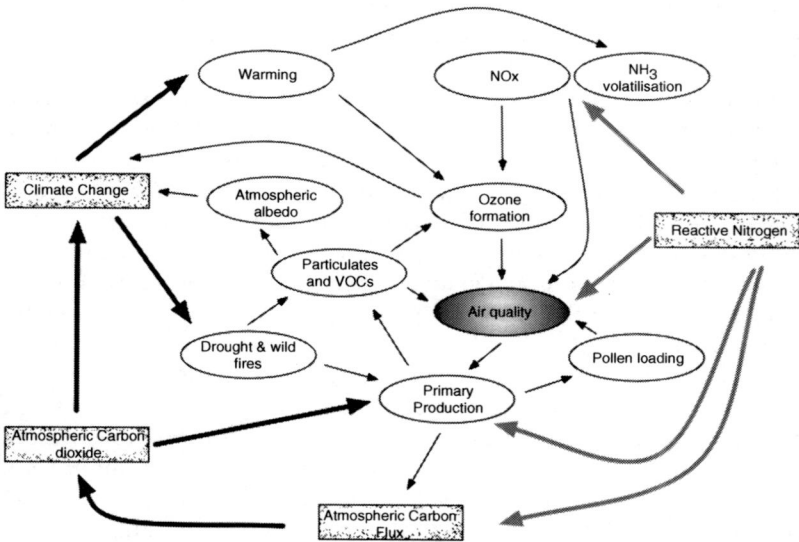

Figure 5.2 The web of nitrogen and climate change interactions in our atmosphere

The major impacts of climate change and increasing carbon dioxide concentrations (large black arrows) and the major impacts of nitrogen (large grey arrows) come together directly in the atmosphere through effects such as ozone formation (from higher temperatures and more NOx emissions).

Source: Dave Reay

In addition to elevating peak temperatures in many cities, climate change is expected to increase the frequency of calm periods, when the air can become stagnant and pollutant concentrations build up more easily[38,39]. Higher temperatures and longer growing seasons for plants may also boost emissions of VOCs, with an overall pattern of longer and more intense ozone episodes expected as climate change intensifies[40]. For instance, applying projected warming and weather pattern changes to high-ozone events in the US through to 2050 indicated that efforts to reduce ozone by tackling NOx emissions will be significantly undermined by the effect of climate change[41] (Figure 5.2).

Climate forcing effects

NOx and ammonia do not act as direct agents of climate forcing. Unlike nitrous oxide, their molecules do not intercept the short-wave energy from the sun or the long-wave heat radiation from the Earth and the lower atmosphere. Instead, their role is as indirect drivers of climate change through the range of atmospheric processes in which they become involved.[42]

As a product of nitrous oxide decomposition, NOx plays an intermediary role in the depletion of stratospheric ozone and so in the climate change impacts that result.

In the lower atmosphere (the troposphere) NOx emissions have a more direct effect on ozone concentrations and therefore on climate forcing. Here, ozone acts as a powerful greenhouse gas[43] with NOx – especially NO_2 – providing the important precursors to its formation and making it an indirect greenhouse gas. This ozone greenhouse effect – around +3 watts per square metre (W m^{-2}) – would make NOx emissions a major source of additional global warming were it not for its domino-like impact on other chemical pathways in the lower atmosphere (Figure 5.3).

On a day-to-day basis, tropospheric ozone formation is often limited by how much NOx is available. Increasing NOx emissions can therefore mean more ozone is produced, and so more warming occurs through the extra heat absorbed and emitted. However, more NOx and ozone production also leads to more production of hydroxyl radicals – the highly reactive molecules that are the atmosphere's primary mechanism for destroying air pollutants. Over longer time periods of months and years, this increase in hydroxyl radicals means that the lifetimes of other atmospheric gases, such as methane, are gradually reduced. As methane is a powerful greenhouse gas, its more rapid removal leads to an overall cooling effect (around –4.5 W m^{-2}), so offsetting some or all of the shorter-term warming caused by the creation of ozone[44]. Whether NOx causes warming or cooling through its impacts on tropospheric ozone appears to depend on where and when it is emitted, with NOx emitted from the surface tending to have an overall cooling effect on the climate and NOx emitted by aircraft having an overall warming effect[45] (Box 5.4). Globally, NOx emissions are now estimated to have a net cooling effect of around –1 W m^{-2}[46].

Both NOx and ammonia emissions can play a further role in net global cooling through their contribution to the formation of small particles (aerosols) in the atmosphere, such as ammonium nitrate. Where these particles form in the size range of 0.1 to 1 microns they are able to effectively reflect and scatter the short-wave energy of sunlight and so increase the overall reflectivity of the atmosphere. The global effect of this airborne nitrogen-induced albedo effect is a small amount of net cooling (around –0.1 W m^{-2}), with additional cooling possible through the enhancing effects of nitrogen aerosols on the reflectivity and lifetime of clouds[46].

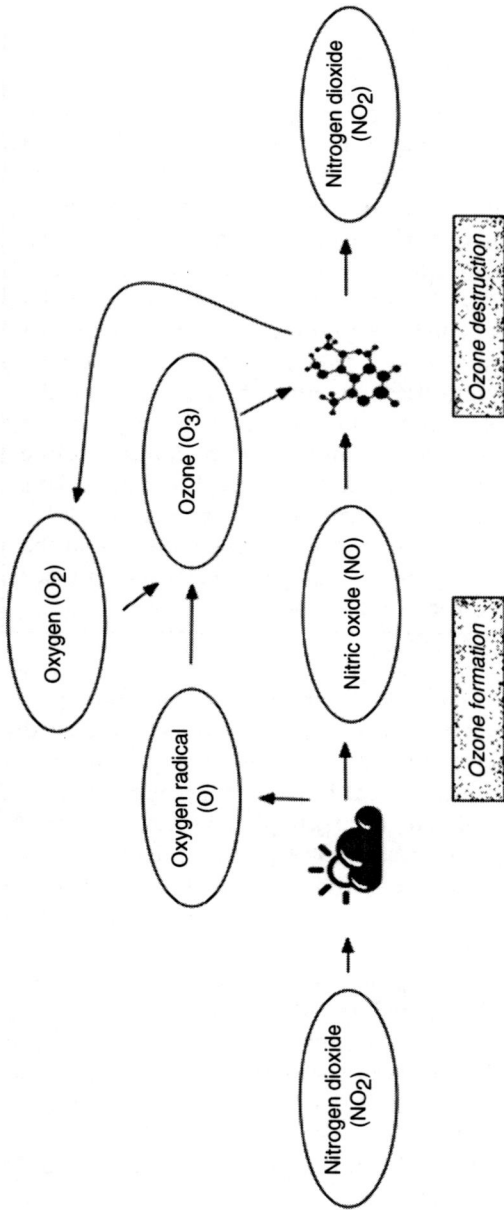

Figure 5.3   Nitrogen oxides and ozone

The boxes show stages at which ozone is either produced or destroyed and the ovals show the forms of nitrogen and oxygen that interact. Ozone formation and destruction is also highly dependent on temperature, sunlight and the supply of VOCs.

Source: Dave Reay

Box 5.4

NOx and aviation's climate impact

Aviation is well known as one of the fastest growing sources of global greenhouse gas emissions[47]. In the UK it still only comprises around 5 per cent of total greenhouse emissions, and globally its carbon dioxide emissions represent just 2.2 per cent of anthropogenic emissions from fossil fuel burning[48]. However, this figure hides the fact that aviation emits a host of atmospheric pollutants as well as carbon dioxide. To account for these other emissions and their climate-forcing impact, something called the 'uplift factor' is used – a factor by which the carbon dioxide emissions are multiplied to estimate the overall climate-forcing effect[49]. Emissions of NOx are especially important for this uplift factor due to their role as an ozone precursor and because, at the high altitudes at which they are emitted, ozone formation is much more efficient than on the Earth's surface. This NOx-derived ozone addition from aviation helps to nearly double the climate uplift factor for aviation emissions, which means that the climate-warming effect of every tonne of fuel burned is almost twice that of the effect from the carbon dioxide alone[50].

The final group of indirect impacts of airborne nitrogen on climate change centres on the interaction of nitrogen with vegetation on the ground, and the altered carbon fluxes that can then ensue. NOx-driven increases in low-level ozone are a major impact of this type, with elevated ozone concentrations being damaging to plant growth and carbon storage[51-53]. By reducing the amounts of photosynthesis-derived carbon stored above and below ground in terrestrial ecosystems, the impacts of tropospheric ozone in Europe alone are estimated to be cutting carbon sequestration by up to 14 million tonnes of carbon each year[46]. As a result of increased ozone, primary production is projected to fall globally by 23 per cent by 2100 (compared to 1901)[52] (Figure 5.4).

The scattering of sunlight by nitrogen-derived aerosols is also thought to affect primary production. The diffuse solar radiation that such scattering causes allows more efficient photosynthesis[54] and so, potentially, greater carbon uptake and storage. However, the weightiest consequences of airborne nitrogen on global carbon stocks and sinks arise when the nitrogen is redeposited to the Earth's surface. It is to such powerful interactions of terrestrial nitrogen with climate change that we turn to in Chapter 6.

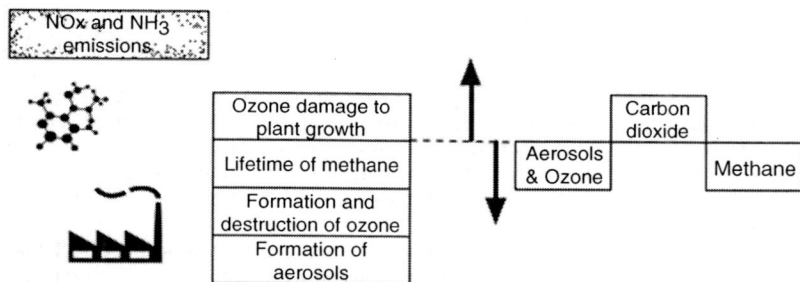

Figure 5.4 NOx and ammonia in the greenhouse effect

For 'NOx and NH$_3$' in the atmosphere, the main effects on greenhouse gases are shown in terms of whether they enhance warming (upward arrow) or reduce it (downward arrow).

Source: Dave Reay

References

1. Delmas, R., Serca, D. & Jambert, C. Global inventory of NOx sources. *Nutrient Cycling in Agroecosystems* **48**, 51–60 (1997).
2. Jaeglé, L., Steinberger, L., Martin, R. V. & Chance, K. Global partitioning of NOx sources using satellite observations: relative roles of fossil fuel combustion, biomass burning and soil emissions. *Faraday Discussions* **130**, 407–423 (2005).
3. Kambara, S., Takarada, T., Toyoshima, M. & Kato, K. Relation between functional forms of coal nitrogen and NOx emissions from pulverized coal combustion. *Fuel* **74**, 1247–1253 (1995).
4. Dignon, J. & Hameed, S. Global emissions of nitrogen and sulfur oxides from 1860 to 1980. *JAPCA* **39**, 180–186 (1989).
5. Galloway, J. N. The global nitrogen cycle: past, present and future. *Science in China. Series C, Life Sciences/Chinese Academy of Sciences* **48 Suppl. 2**, 669–678, doi:10.1007/BF03187108 (2005).
6. Davidson, E. A. & Kingerlee, W. A global inventory of nitric oxide emissions from soils. *Nutrient Cycling in Agroecosystems* **48**, 37–50 (1997).
7. Yienger, J. & Levy, H. Empirical model of global soil-biogenic NOx emissions. *Journal of Geophysical Research: Atmospheres (1984–2012)* **100**, 11447–11464 (1995).
8. Jain, A. K., Tao, Z., Yang, X. & Gillespie, C. Estimates of global biomass burning emissions for reactive greenhouse gases (CO, NMHCs, and NOx) and CO$_2$. *Journal of Geophysical Research: Atmospheres (1984–2012)* **111**, D06304 (2006).
9. Samet, J. & Utell, M. The risk of nitrogen dioxide: what have we learned from epidemiological and clinical studies? *Toxicology and Industrial Health* **6**, 247–262 (1990).
10. Dales, R., Burnett, R. T., Smith-Doiron, M., Stieb, D. M. & Brook, J. R. Air pollution and sudden infant death syndrome. *Pediatrics* **113**, e628–e631 (2004).
11. Sillman, S., Logan, J. A. & Wofsy, S. C. The sensitivity of ozone to nitrogen oxides and hydrocarbons in regional ozone episodes. *Journal of Geophysical Research: Atmospheres (1984–2012)* **95**, 1837–1851 (1990).

12. Koren, H. S. et al. Ozone-induced inflammation in the lower airways of human subjects. *American Review of Respiratory Disease* **139**, 407–415 (1989).

13. Bell, M. L., McDermott, A., Zeger, S. L., Samet, J. M. & Dominici, F. Ozone and short-term mortality in 95 US urban communities, 1987–2000. *JAMA: The Journal of the American Medical Association* **292**, 2372–2378 (2004).

14. Harrison, R. M. & Yin, J. Particulate matter in the atmosphere: which particle properties are important for its effects on health? *Science of the Total Environment* **249**, 85–101 (2000).

15. Bell, M. L. & Davis, D. L. Reassessment of the lethal London fog of 1952: novel indicators of acute and chronic consequences of acute exposure to air pollution. *Environmental Health Perspectives* **109**, 389 (2001).

16. Sillman, S. & Samson, P. J. Impact of temperature on oxidant photochemistry in urban, polluted rural and remote environments. *Journal of Geophysical Research: Atmospheres (1984–2012)* **100**, 11497–11508 (1995).

17. Loomis, D., Castillejos, M., Gold, D. R., McDonnell, W. & Borja-Aburto, V. H. Air pollution and infant mortality in Mexico City. *Epidemiology* **10**, 118–123 (1999).

18. Finlayson-Pitts, B. & Pitts Jr, J. Atmospheric chemistry of tropospheric ozone formation: scientific and regulatory implications. *Air & Waste* **43**, 1091–1100 (1993).

19. Kesselmeier, J. & Staudt, M. Biogenic volatile organic compounds (VOC): an overview on emission, physiology and ecology. *Journal of Atmospheric Chemistry* **33**, 23–88 (1999).

20. Zolghadri, A., Monsion, M., Henry, D., Marchionini, C. & Petrique, O. Development of an operational model-based warning system for tropospheric ozone concentrations in Bordeaux, France. *Environmental Modelling & Software* **19**, 369–382 (2004).

21. Farrell, A., Carter, R. & Raufer, R. The NOx budget: market-based control of tropospheric ozone in the northeastern United States. *Resource and Energy Economics* **21**, 103–124 (1999).

22. Reay, D. S., Dentener, F., Smith, P., Grace, J. & Feely, R. A. Global nitrogen deposition and carbon sinks. *Nature Geoscience* **1**, 430–437, doi:10.1038/ngeo230 (2008).

23. Burtraw, D., Palmer, K., Bharvirkar, R. & Paul, A. Cost-effective reduction of NOx emissions from electricity generation. *Journal of the Air & Waste Management Association* **51**, 1476–1489 (2001).

24. ApSimon, H. M. & Warren, R. F. Transboundary air pollution in Europe. *Energy Policy* **24**, 631–640 (1996).

25. Lawrence, M. G. & Crutzen, P. J. Influence of NOx emissions from ships on tropospheric photochemistry and climate. *Nature* **402**, 167–170 (1999).

26. Derwent, R. & Friedl, R. Impacts of aircraft emissions on atmospheric ozone. In *IPCC Special Report on Aviation and the Global Atmosphere*, edited by J. E. Penner et al., 29–64 (Cambridge University Press, UK, 1999).

27. Stevenson, D. S. & Derwent, R. G. Does the location of aircraft nitrogen oxide emissions affect their climate impact? *Geophysical Research Letters* **36**, L17810 (2009).

28. Holland, E. A., Dentener, F. J., Braswell, B. H. & Sulzman, J. M. Contemporary and pre-industrial global reactive nitrogen budgets. In *New Perspectives*

on Nitrogen Cycling in the Temperate and Tropical Americas, edited by A. R. Townsend, 7–43 (Springer, 1999).

29. Vitousek, P. M. et al. Human alteration of the global nitrogen cycle: sources and consequences. *Ecological Applications* **7**, 737–750 (1997).

30. Sutton, M., Dragosits, U., Tang, Y. & Fowler, D. Ammonia emissions from non-agricultural sources in the UK. *Atmospheric Environment* **34**, 855–869 (2000).

31. Schwartz, J., Laden, F. & Zanobetti, A. The concentration-response relation between PM (2.5) and daily deaths. *Environmental Health Perspectives* **110**, 1025 (2002).

32. Haeuber, R., Peel, J., Garcia, V., Neas, L. & Russell, A. Implications of nitrogen-climate interactions for ambient air pollution and human health. In *AGU Fall Meeting Abstracts*, 4 pp., 2011AGUFM.B42C-04 (American Geophysical Union, USA, 2011).

33. Knowlton, K. et al. Assessing ozone-related health impacts under a changing climate. *Environmental Health Perspectives* **112**, 1557–1563 (2004).

34. Zeng, G., Pyle, J. & Young, P. Impact of climate change on tropospheric ozone and its global budgets. *Atmospheric Chemistry and Physics* **8**, 369–387 (2008).

35. Taha, H. Urban climates and heat islands: albedo, evapotranspiration, and anthropogenic heat. *Energy and Buildings* **25**, 99–103 (1997).

36. Rosenfeld, A. H., Akbari, H., Romm, J. J. & Pomerantz, M. Cool communities: strategies for heat island mitigation and smog reduction. *Energy and Buildings* **28**, 51–62 (1998).

37. Benjamin, M. T. & Winer, A. M. Estimating the ozone-forming potential of urban trees and shrubs. *Atmospheric Environment* **32**, 53–68 (1998).

38. Suddick, E. C., Whitney, P., Townsend, A. R. & Davidson, E. A. The role of nitrogen in climate change and the impacts of nitrogen–climate interactions in the United States: foreword to thematic issue. *Biogeochemistry* **114**, 1–10 (2013).

39. Jacob, D. J. & Winner, D. A. Effect of climate change on air quality. *Atmospheric Environment* **43**, 51–63 (2009).

40. Wilby, R. L. Constructing climate change scenarios of urban heat island intensity and air quality. *Environment and Planning. B, Planning & Design* **35**, 902 (2008).

41. Wu, S. et al. Effects of 2000–2050 global change on ozone air quality in the United States. *Journal of Geophysical Research: Atmospheres (1984–2012)* **113**, D06302 (2008).

42. Seinfeld, J. H. & Pandis, S. N. *Atmospheric chemistry and physics: from air pollution to climate change.* (John Wiley & Sons, 2012).

43. Solomon, S. *Climate change 2007 – the physical science basis: Working Group I contribution to the fourth assessment report of the IPCC.* Vol. 4. (Cambridge University Press, 2007).

44. Wild, O., Prather, M. J. & Akimoto, H. Indirect long-term global radiative cooling from NOx emissions. *Geophysical Research Letters* **28**, 1719–1722 (2001).

45. Derwent, R. et al. Radiative forcing from surface NOx emissions: spatial and seasonal variations. *Climatic Change* **88**, 385–401 (2008).

46. Butterbach-Bahl, K. et al. Nitrogen as a threat to the European greenhouse balance. In *The European Nitrogen Assessment: Sources, Effects and Policy*

Perspectives, edited by M. A. Sutton et al., chap. 19, 434–462 (Cambridge University Press, UK, 2011).

47. Ellis, J. & Treanton, K. Recent trends in energy-related CO_2 emissions. *Energy Policy* 26, 159–166 (1998).
48. Macintosh, A. & Wallace, L. International aviation emissions to 2025: can emissions be stabilised without restricting demand? *Energy Policy* 37, 264–273 (2009).
49. Bows, A. & Anderson, K. L. Policy clash: can projected aviation growth be reconciled with the UK Government's 60% carbon-reduction target? *Transport Policy* 14, 103–110 (2007).
50. Bows, A., Upham, P. & Anderson, K. Growth Scenarios for EU & UK Aviation: contradictions with climate policy. *Tyndall Centre for Climate Change/Friends of the Earth, Manchester* (2005).
51. Felzer, B. S. et al. Past and future effects of ozone on net primary production and carbon sequestration using a global biogeochemical model. In *MIT Joint Program on the Science and Policy of Global Change*. Report no. 103, http://hdl.handle.net/1721.1/4053 (2003).
52. Sitch, S., Cox, P., Collins, W. & Huntingford, C. Indirect radiative forcing of climate change through ozone effects on the land-carbon sink. *Nature* 448, 791–794 (2007).
53. Felzer, B. et al. Future effects of ozone on carbon sequestration and climate change policy using a global biogeochemical model. *Climatic Change* 73, 345–373 (2005).
54. Mercado, L. M. et al. Impact of changes in diffuse radiation on the global land carbon sink. *Nature* 458, 1014–1017 (2009).

6
Terrestrial Nitrogen and Climate Change

In terms of global climate, a leading impact of changing nitrogen input to terrestrial ecosystems is the huge alteration in nitrous oxide emissions it induces, especially in agricultural soils (see Chapter 4). As inputs have spiralled upwards over the last century, emissions of nitrous oxide via denitrification and nitrification in soils have followed suit. Globally, emissions from agricultural activities are now estimated to be between five and seven million tonnes of nitrogen as nitrous oxide each year[1]. The bulk of this comes directly from the nitrogen-enriched soils themselves, with the remainder either being emitted from manure stores or arising from the reactive nitrogen that is lost from the soils due to leaching or volatilisation. Nitrous oxide has a global warming potential (GWP) of around 300 times that of carbon dioxide on a mass basis – GWP is a way of standardising the amount of warming caused by different greenhouse gases so they can be compared to carbon dioxide[2]. Annual nitrous oxide emissions from agriculture are therefore equivalent to enhancing the greenhouse effect to the tune of about three billion extra tonnes of carbon dioxide a year – the annual emissions of a billion cars.

As efforts to tackle anthropogenic climate change become more aggressive in the 21st century, addressing nitrous oxide emissions from land use is set to become a key battleground. Rapidly rising demand for food and biofuels means increased pressure to cultivate new land and intensify production in existing managed areas. Enormous increases in nitrogen inputs to agricultural systems may be needed to power this global boom in food, fuel and fibre production. This could then enhance nitrous oxide emissions to a level where they completely undermine climate change mitigation efforts in other sectors[1]. Making nitrogen inputs to terrestrial ecosystems part of the solution instead of part of the problem will require a step change in the average efficiency of its use in agriculture and call for the uptake of strategies that can deliver reduced nitrous oxide emissions.

Each of the various climate change scenarios for the 21st century is highly contingent on the emissions pathway we follow in the next decade or so. Here lies the greatest uncertainty in projecting future warming: which emissions pathway will global society choose to go down? If we fail to address nitrogen's growing role as a driver of greenhouse gas fluxes, we increasingly load the dice in favour of the pathway to dangerous climate change.

Despite the inequalities and inefficiencies of its use, the ammonia-producing discoveries of Haber and Bosch (Chapter 2) have shored up massive and ongoing human population growth. It has been estimated that without their process agriculture could support only about half of today's human population, and then only on the Spartan low-meat diets of the 1900s[3,4]. The process has become a cornerstone of global food supply, and doing without it is no more an option for 21st-century humanity than is abandoning antibiotics. Like antibiotics though, the true price to pay for our poor use of this magnificent tool of civilisation has been slow to emerge. It was in the patchwork fields of John Lawes' 19th-century Broadbalk experiment[5] that the first ecosystem canaries – plant species stifled by a spreading plume of reactive nitrogen – gave warning of what lay ahead.

Most plants have had to evolve in a world with a vanishingly short supply of available nitrogen. For millennia they have fought a daily battle for whatever became available from decomposition, leaky legumes or occasional showers of lightning-enriched rain and snow. As human activities began to add more and more to the global flow, some species were able to feast and prosper on the enriched diet, while others were literally smothered.

Lawes and his successors systematically noted changes in biodiversity as well as the yields of croplands and pasture at Broadbalk. Year after year some plots received extra reactive nitrogen, either in the form of manure or as chemical fertiliser, while others were left alone. After many decades more than 50 different plant species continued to thrive in the buzzing tangle of the unfertilised plots, but the number of species had dwindled to fewer than 10 in the areas receiving heavy doses of nitrogen[3]. Even for those plants able to lap up the extra helpings of reactive nitrogen, the lush growth that it engenders can also be a boon for pests and diseases, turning tough, well-protected plant foliage into a bloated food store ripe for attack[6,7]. Such negative side effects of nitrogen enrichment would be more acceptable if they occurred only where it was applied. If the world's croplands and pastures ended up as power-houses of plant growth dominated by just one or two species, then this could be offset by leaving other areas alone – havens for the native, low-nutrient-loving species. The big problem for biodiversity and climate alike is that reactive nitrogen rarely stays where it is put, with the agricultural nitrogen lost to the air as nitrogen oxides (NOx) and ammonia (NH_3) emissions joining that from industry and fossil fuel burning to create an intensifying shower of nitrogen enrichment over the world's terrestrial ecosystems.

The terrestrial nitrogen budget

There are an estimated 130–140 billion tonnes (Pg) of nitrogen contained in the top metre of the world's soils, compared to only about 10 Pg contained in plants[8]. Nitrogen inputs to managed ecosystems around the world are now dominated by the addition of nitrogen fertilisers, manures and the cultivation of legumes in the agricultural sector, with additional inputs to both managed and natural land areas via atmospheric deposition and natural nitrogen fixation[9] (Figure 6.1).

While application of synthetic nitrogen fertilisers to agricultural lands – courtesy of Haber and Bosch – has only boomed in the post-war period, large nitrogen additions via manure and the planting of legumes date back more than 10,000 years. Across much of the developed world synthetic fertiliser addition is now the largest input of nitrogen to managed lands, closely followed by that contained in manure applications. The act of converting unmanaged land for agricultural use can itself boost the short-term availability of nitrogen, as the organic nitrogen buried in the soils is exposed and mineralised to ammonium (called 'ammonification'). In the longer term the continual planting, ploughing and re-exposure of cultivated soils effectively runs down their natural nitrogen supply[10] and larger deliberate inputs via fertilisers and manure are required.

Like synthetic fertilisers, the surge in airborne nitrogen inputs to terrestrial systems has been relatively recent, with the industrial revolution marking the point at which emissions and deposition began to climb fast. Alongside industrialisation and rising NOx emissions from fossil fuel burning, the intensification of agriculture and associated ammonia emissions has led to a three- to fivefold increase in reactive nitrogen emissions to the atmosphere over the last century[2]. Global emissions are mainly terrestrial in origin, and stand at over 100 million tonnes of nitrogen a year. More than half of all these emissions are redeposited (Box 6.1) to terrestrial ecosystems[11].

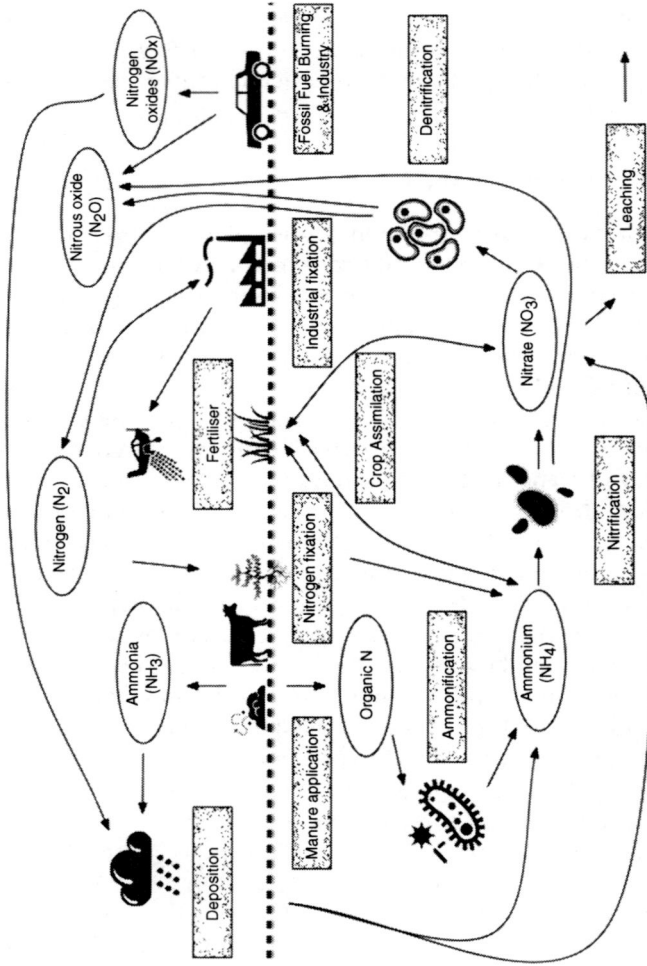

Figure 6.1 The terrestrial nitrogen cycle

The dotted line represents the land surface with the boxes showing the key processes by which nitrogen is cycled and the ovals showing the changing form of nitrogen as it undergoes these processes. Human impacts on the terrestrial nitrogen cycle are dominated by our direct addition of nitrogen fertilisers to soils, by nitrogen fixation by legumes, and by the deposition of reactive nitrogen produced by farming and fossil fuel burning.

Source: Dave Reay

Climate forcing and carbon fluxes

In addition to its direct impact on climate forcing through nitrous oxide emissions, reactive nitrogen in terrestrial ecosystems plays a key indirect role in determining the magnitude of the vegetation and soil carbon sinks and of land-based methane fluxes (Figure 6.3).

As concentrations of carbon dioxide in our atmosphere continue to rise, vegetation in many areas of the world is expected to respond by growing faster, which is called the 'CO_2 fertilisation effect'. On a large enough scale this expansion in carbon uptake by vegetation could help to buffer emissions from fossil fuel burning and buy us time in the fight to avoid dangerous climate change. However, many of the plants that would in theory grow better in this higher CO_2 world are in fact limited by other factors such as water supply, pests and, very importantly, nitrogen. For instance, including the nitrogen cycle in climate projections through 2100 suggests that the amount of extra uptake of carbon dioxide on land that should occur, due to the plants benefiting from the CO_2 fertilisation effect, is slashed by almost 75 per cent[16].

Box 6.1

Nitrogen deposition

Total nitrogen deposition from the atmosphere to the land and oceans takes two forms: wet deposition and dry deposition (Figure 6.2). Dry deposition commonly refers to the deposition rate of nitrogen in the form of a gas or as an aerosol (e.g. nitrates)[12]. The rate of such dry deposition depends on how the nitrogen-rich gas or aerosol interacts with the surface – forested areas are often able to scavenge reactive nitrogen from the air much more effectively than bare soils, with the leaves of the trees providing a large surface area onto which the nitrogen is collected[13]. For larger nitrogen-rich aerosols, dry deposition can also occur via the particles landing directly onto the surface, with wind speed and the type of surface again being crucial in determining just how much is deposited[14].

Wet deposition involves the scavenging of reactive nitrogen in the air by droplets of water, which are then deposited to the land or oceans in snow, rain or mist. A common example of wet deposition is the formation of NOx gases by lightning, with the gases then being converted into nitrate and deposited to the land or oceans through rainfall[15].

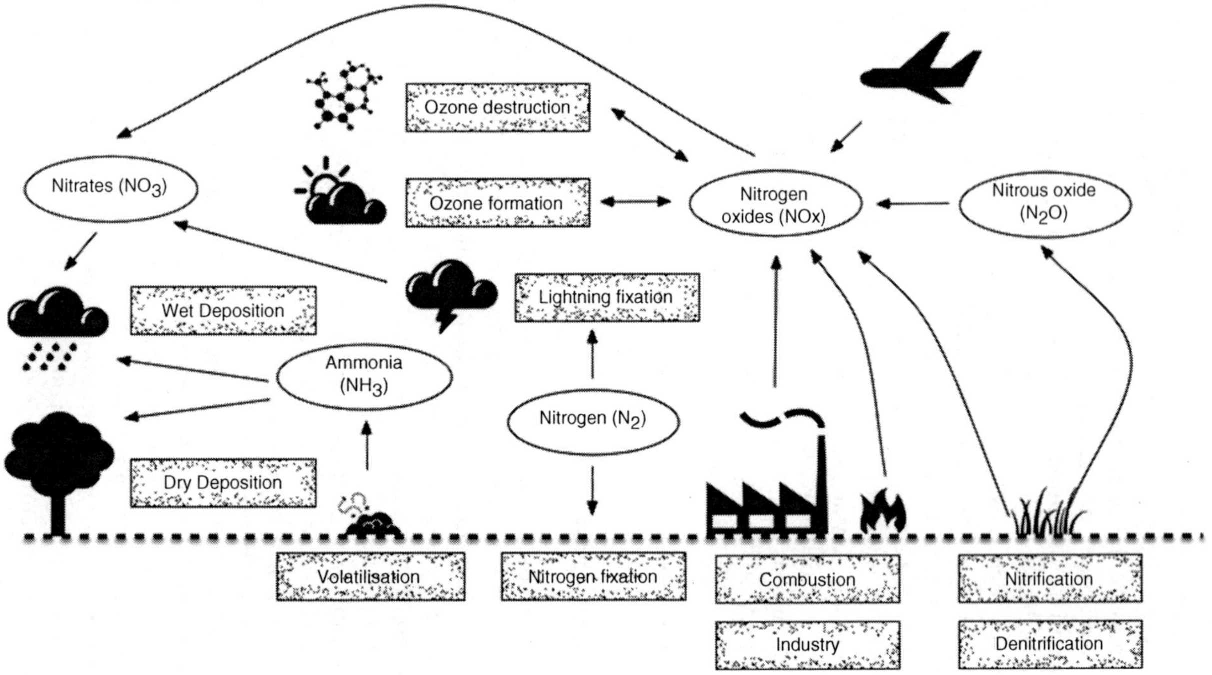

Figure 6.2   Nitrogen in the atmosphere and its deposition

The dotted line represents the land surface with the boxes showing the key processes by which nitrogen is cycled and the ovals showing the changing form of nitrogen as it undergoes these processes. Reactive nitrogen is deposited via either wet deposition (in rain, snow and mist) or dry deposition (in particles or as aerosols landing on vegetation, soils or water).

Source: Dave Reay

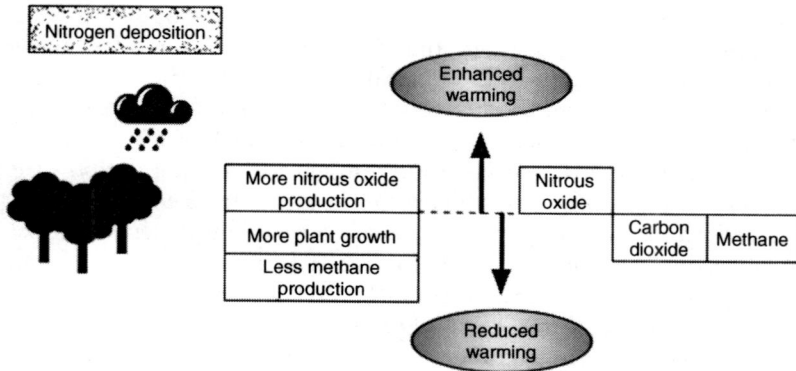

Figure 6.3 Nitrogen deposition and the greenhouse effect

For reactive nitrogen deposition to vegetation and soils, the main effects on greenhouse gases are shown in terms of whether they enhance warming (upward arrow) or reduce it (downward arrow).

Source: Dave Reay

To accurately project future carbon fluxes around the world, we therefore require a much better understanding of how these fluxes will interact with changing nitrogen inputs. Unfortunately, gaining this real depth of understanding remains a challenge due to the large number of ways in which changing inputs can affect carbon cycling in the numerous different ecosystems. As these responses can vary over scales of single cells to continents and over times of milliseconds to millennia, to get even a broad indication of whether more reactive nitrogen will increase or decrease the global carbon sinks requires a look at studies covering all the major ecosystem types on the planet[11].

Nitrogen and the forest carbon sink

In the 1980s the global terrestrial carbon sink strength stood at around one-third of a billion tonnes per year (0.3 Pg C y^{-1}). By the 1990s this had risen to one billion tonnes (1.0 Pg C y^{-1})[2]. Much of this large terrestrial carbon sink seemed to be associated with northern temperate forests growing between about 25° and 55°N. The suggested reasons for this major expansion in the forest carbon sink included forest regrowth, the CO_2 fertilisation effect, changes in land use and climate change[17], yet throughout this period these forests were also experiencing an intensifying rain of reactive nitrogen. Quite small stimulations of the growth rate, death rate or decomposition rate in forest ecosystems can cause a large change in the carbon sink, because the total carbon pool in these eco-systems is huge. The possibility that reactive nitrogen was a key player in increasing forest growth rates became evident with the finding that many northern and temperate forests (but not tropical rain forests) are nitrogen-limited[18]. A nagging question has therefore been whether the large late-20th century increase in terrestrial carbon sequestration – the so-called 'missing sink' – was linked to increasing nitrogen deposition.

The circumstantial evidence was strong, with a simultaneous increase in nitrogen deposition to northern forests and their carbon content having been observed since the 1950s[19]. Yet reports of limited incor-poration of added nitrogen and high losses to drainage water in other northern forests suggested that elevated nitrogen deposition had played only a minor role in boosting the size of the northern forest carbon sink[20]. Then, in 2008, a controversial study in the leading scientific journal *Nature* reported that net carbon sequestration in a range of tem-perate and boreal forests had indeed responded to elevated nitrogen deposition resulting from human activities[21]. The effect in the forests studied was very large, always positive and demonstrated the indirect control of forest carbon fluxes exerted by human use of nitrogen – it was inferred that as much as 200 grams of carbon might be seques-tered for every gram of reactive nitrogen added. Though some forests in these temperate regions were likely to be nitrogen-saturated and there-fore unresponsive to more nitrogen deposition, it seemed that the bulk of forests in this global engine room of terrestrial carbon storage had responded strongly in the past to more reactive nitrogen and might therefore do so in the future.

Despite greater controls on reactive nitrogen emissions in a number of countries, there seems little doubt that northern forests will continue to receive an increasing supply of reactive nitrogen from the air over the next few decades[22]. Assuming that key factors such as forest distribution and age class structure remain constant, using the nitrogen-induced boost to forest carbon uptake reported in the controversial 2008 *Nature* study[21], one might expect to see an additional 300 million tonnes of carbon (0.3 Pg C) uptake per year in European forests alone. This would represent a huge increase, equivalent to them soaking up the annual human-induced greenhouse gas emissions from the whole of the UK. In North America, with a much larger forest area (about 770 million hectares) than Europe, such a nitrogen response would equate to sequestration of an additional 1.5 billion tonnes of carbon (1.5 Pg C) each year – a really substantial buffer to growing carbon dioxide emis-sions from fossil burning around the world.

Large increases in reactive nitrogen deposition to northern forests over the next few decades could therefore help to sequester one-quarter of annual anthropogenic carbon dioxide emissions and thereby significantly reduce the growth rate of atmospheric concentrations. Unfortunately, the potential flaws in estimation of this nitrogen-enriched lid on human-induced climate change are legion[11].

Firstly, it is very likely that additional controls on emissions of air pollutants will be put in place by many nations in the next 20 years or so, making large increases in nitrogen deposition to northern forests unrealistic[22]. It is also a big assumption that the forest age class structure in 2030 will be identical to that of today, as mature forests have shown only a limited response to elevated nitrogen inputs and an ageing forest stock could be less responsive to more nitrogen. Most crucially, there is the extrapolation of the sizeable response of forest carbon sequestration to elevated nitrogen deposition given in the controversial *Nature* study to all European and North American forests. Some of these forests are likely to already be nitrogen-replete or to become so in the near future, while other studies have reported much smaller responses to elevated nitrogen deposition in these regions[23]. Overall then, even if there is a substantial increase in nitrogen deposition rate by 2030, the consequent increase in carbon sequestration by northern forests would likely be relatively minor.

Nitrogen and the soil carbon sink

Evidence for changes in soil carbon sinks due to reactive nitrogen enrichment comes from a variety of sources, including changes in leaf litter decomposition rates and the soil organic carbon stock. However, much of this evidence is contradictory; some studies suggest that soil carbon may decrease under nitrogen enrichment, others suggest no change, and yet others suggest that soil carbon sinks may increase. The net effect of more nitrogen inputs on the global soil carbon sink therefore remains unclear. The response will depend upon the balance between the nitrogen-induced increases in carbon inputs to the soil – through increased plant growth as discussed earlier – and the influence of increased reactive nitrogen on carbon losses via soil organic carbon decomposition, respiration and mineralisation.

In agricultural soils, nitrogen fertilisation can enhance organic carbon mineralisation, but studies of soil respiration suggest no change. Indeed, mineralisation has been shown to be retarded at very high nitrogen concentrations and, in long-term experiments in agricultural systems, artificial nitrogen fertilisation at much higher rates than those received from natural deposition has reportedly led to some small increases in soil carbon[11].

In forest soils too, the evidence is contradictory. Increases in soil respiration (i.e. short-term carbon loss) in response to nitrogen fertilisation have been reported. It has also been suggested that relatively low rates of nitrogen addition can suppress soil respiration. As such, some soil carbon stocks may therefore increase as a consequence of increased nitrogen deposition in the future, but the saturation of this response remains unexplored.

Increasing nitrogen inputs, it seems, do not provide a free ride to greatly elevated forest or soil carbon sequestration. Even if a significant increase in the size of the terrestrial carbon sink were a beneficial side effect of burgeoning nitrogen use and emissions, its help in addressing climate change would have to be balanced against any enhancement of nitrous oxide emissions. While a doubling of reactive nitrogen emissions by 2030 could boost carbon dioxide uptake in forests by as much as three billion tonnes a year, the nitrous oxide penalty would offset up to 90 per cent of this nitrogen pollution silver lining[11].

In coming decades the protection of the existing terrestrial carbon sinks from deforestation and land-use change should be the focus of carbon sink management. Yes, there is some potential for enhancing the vegetation and soil carbon sinks through nitrogen addition, but it is the thinnest of veils to cover the many sins of reactive nitrogen emission. Relying on yet more air, water and soil pollution to help address human-induced climate change is reminiscent of the release of cane toads to tackle the cane beetle in Australia: an ineffective solution that creates a whole new set of problems.

Nitrogen and terrestrial methane fluxes

Just as for forest and soil carbon sinks, the precise impact of changing nitrogen inputs on terrestrial methane fluxes remains highly uncertain, with much of this uncertainty stemming from the myriad ways in which nitrogen interacts with the process of methane production (called 'methanogenesis') and consumption (methane oxidation). Where increasing nitrogen availability boosts plant growth and organic matter inputs to soils, there is the potential for an increase in methane production and emissions from major terrestrial methane sources such as wetlands and rice paddies[24]. In the low- and zero-oxygen environments of these water-logged soils, however, higher concentrations of nitrate can favour denitrifying microbes at the cost of the methane producers (methanogens), thereby cutting methane emissions[25]. In most natural wetlands nitrate concentrations are too low for this methanogen inhibition to be very strong, but in wetlands constructed for wastewater treatment and in fertilised rice paddy soils the role of nitrogen in limiting methane production may be substantial[26].

Other terrestrial ecosystems, such as well-drained forest soils, can act as effective sinks for atmospheric methane. Globally these soils represent a sink for methane of around 30 million tonnes each year[2] and here again nitrogen can play an important role in determining the net flux of methane. In well-aerated soil methane is predominantly used by aerobic microbes (called 'methanotrophs'), which use it as a source of carbon in the process of biological methane oxidation. As well as the water content of the soil, factors such as pH, soil temperature and the concentration of inorganic nitrogen (e.g. nitrate, nitrite and ammonium) can be crucial in determining whether or not a particular soil will act as a sink for methane[27]. For instance, though the soil under alder trees may be well-aerated and seemingly ideal for methanotrophs, the high concentrations of nitrate which arise from nitrogen fixation in the alder's roots can mean that methanotroph activity is slowed or completely stopped[28]. Conversion of wooded and fallow land for agricultural use tends to result in increased nitrogen concentrations in the soil, which may then inhibit methane oxidation[29]. Similarly, the increased deposition of nitrogen from the atmosphere due to human activities can also reduce, or completely inhibit, methane oxidation in soils. However, the true picture of how nitrogen interacts with methane production and consumption is complicated by evidence that suggests that where the methanotrophs are nitrogen-limited, reactive nitrogen inputs may actually help stimulate methane consumption rather than reduce it[30]. The amount and form of nitrogen inputs, as well as the type of ecosystem and its conditions, make for a complex web of interactions that determine whether extra nitrogen means more or less methane ends up in the atmosphere.

Climate change impacts

Nitrous oxide feedbacks

The impacts of future climate change on nitrous oxide emissions from terrestrial ecosystems are potentially significant, yet remain difficult to quantify on a global scale. Soil nitrification and denitrification – the main sources of global nitrous oxide emissions – are both highly dependent on soil moisture and temperature. The major changes in precipitation and temperature that constitute climate change could therefore radically increase or decrease soil nitrous oxide emissions in any given area.

A recent examination of modelled nitrous oxide emissions from Australian pasture-based dairy systems under future climate change scenarios indicated an increase in emissions at three of the four sites studied[31]. Conversely, modelling of nitrous oxide emissions from a humid pasture in Ireland under future climate change indicated that a significant increase in above-ground biomass and associated nitrogen demand would serve to avoid significant increases in nitrous oxide emissions[32] – with the extra plant growth intercepting much of the nitrogen before it could be used by the nitrifiers and denitrifiers. The few direct studies of agricultural nitrous oxide fluxes under simulated future climates do suggest increased emissions in response to warming[33], but indicate little effect of summer drought or elevated carbon dioxide concentration on annual fluxes[34]. Overall, it is likely that changes in food demand, land management and nitrogen use efficiency will be much more important determinants of global nitrous oxide emissions than climate change in the 21st century. However, significant indirect effects of climate change on agricultural nitrous oxide fluxes – such as reduced crop and livestock productivity[35,36], altered nitrogen leaching rates[37], and enhanced ammonia volatilisation[38,39] – require further investigation and quantification.

Certainly, those impacts of climate change that lead to a net decrease in food production – such as more drought damage to crops and heat stress in livestock – may reduce nitrogen use efficiency in agriculture and push up overall nitrogen inputs. Likewise, any increase in the intensity of rainfall events risks enhancing the rates of nitrogen leaching and run-off from many agricultural systems. This may exacerbate nitrogen pollution issues in freshwater (Chapter 7) and marine (Chapter 8) systems and, alongside any heatwave-induced increases in ammonia volatilisation, boost indirect nitrous oxide emissions from agriculture globally. Enhanced ammonia volatilisation may be especially important in livestock production, where higher temperatures are the primary driver of ammonia losses from manure and urine[40]. Several studies have also reported that soil warming could accelerate nitrogen mineralisation and therefore increase its availability and turnover rates in soils. This may further enhance soil emissions of nitrous oxide and NOx, as well as nitrogen leaching[41,42].

Climate, nitrogen and plant interactions

Increased nitrogen availability often leads to faster plant growth, but it can also alter the physical characteristics of plant tissues and interact with the timing of key climate-dependent events such as bud burst and flowering. As such, it can increase the susceptibility of plants to damage from drought, fire, frosts and pests[43,44]. Climate change itself is projected to alter the timing, frequency and intensity of these impacts, thereby exacerbating the nitrogen-enrichment effects in some areas. In European heathlands, for instance, larvae of the heather beetle – a significant pest in these ecosystems – grow better on plant shoots with high nitrogen content. As nitrogen inputs increase, heather beetle and other pest attacks can become more and more damaging[45]. Climate change may simultaneously increase winter survival of such pests[46] and enhance the overall risk of plant damage and loss even further (Figure 6.4).

Biodiversity

The impacts of climate change on biodiversity in terrestrial ecosystems are expected to be far-reaching in many areas. Up to a third of land animal and plant species may be committed to extinction by 2050 due to impacts such as habitat loss[47]. Increasing nitrogen inputs pose a major

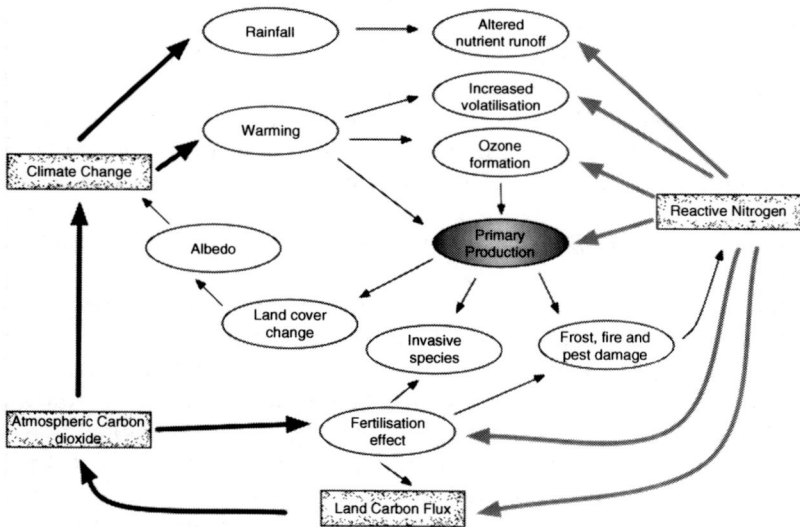

Figure 6.4 Key interactions of nitrogen and climate change on land

The major impacts of climate change and increasing carbon dioxide concentrations (large black arrows) and the major impacts of nitrogen (large grey arrows) come together directly in the terrestrial environment through interactions such as altered vegetation growth (from more nitrogen inputs and increased carbon dioxide). In some cases the extra growth induced by more nitrogen and carbon dioxide actually makes the vegetation more susceptible to damage by fire, pests and frosts.

Source: Dave Reay

additional threat to biodiversity in many terrestrial ecosystems, potentially exacerbating climate and land-use change impacts and causing further species loss[48].

More than one-tenth of the world's natural vegetation now receives in excess of one gram of nitrogen per square metre each year (m^{-2} y^{-1}) as deposition from the atmosphere. The terms 'nitrogen saturation' and 'critical load exceedance' have been used to describe the process whereby negative effects of such excess deposition are observed and leakage of reactive nitrogen from the ecosystem is evident[11]. Reductions in plant

---

Box 6.2

Nitrogen and acidification

Like sulphur dioxide in pea soupers, NOx gases can also be oxidised in the air and combined with water to form nitrates and acids. In cities these acids may attack buildings, gradually dissolving any limestone, marble or sandstone on which they end up[50]. Spreading out into the countryside, the nitrates and acids rain down on vegetation, soils and lakes to slowly but surely wipe out plants and animals unable to adapt to the increasingly acidic and nitrogen-rich conditions. In the 1970s sulphur was estimated to contribute roughly two-thirds of the acidic deposition, with the other third of this acidic input being derived from nitrogen (N) compounds deposited as nitrate in rain and nitrogen dioxide and nitric acid in dry deposition. As sulphur emissions have been reduced, the role of reactive nitrogen in acidifying the myriad environments on which it falls has grown – the decline in sulphur emissions occurring during a period in which airborne nitrogen emissions have increased substantially[51]. Burgeoning NOx emissions from fossil fuel burning, and the acidic rainfall that they contribute to, have played a part, but more devastating to many ecosystems has been the sucker punch of ammonia. Deposited in ever-greater quantities as humankind's hunger for meat and dairy produce intensifies, the ammonia so gladly received by the nitrifying bacteria in the soil is used to generate the stream of hydrogen ions that push the pH lower and lower[52]. Acidified ecosystems are not only bad for many of the plants and animals that live in them, they also reduce the ability of the soils to hold onto toxic heavy metals, allowing increasing amounts to leach out into drainage waters and ultimately into the drinking water supplies that many millions rely on.

diversity in response to increasing inputs have been recorded widely since Lawes' first observations at Broadbalk in the 19th century. These losses are often as a result of the additional nitrogen supply allowing fast-growing species to invade new areas and to outcompete less nitrogen-loving plants[49]. Biodiversity losses may also occur due to the greater risks of frost, fire, drought and pest damage, as discussed above. A major additional threat posed by increased reactive nitrogen inputs is that of soil acidification[43], whereby the more acid soil conditions induced by nitrogen inputs favour more acid-loving plant species at the expense of other more alkaline or neutral pH-loving plants (Box 6.2).

The impacts of increased nitrogen inputs on terrestrial ecosystems and their interactions with climate change remain a crucial area for research and a considerable source of uncertainty in projections. In addition to their importance to global carbon stocks and greenhouse gas fluxes, they have major implications for land management strategies, species conservation efforts and the regulation of nitrogen emissions. The challenge of reducing terrestrial nitrous oxide emissions while still feeding an expanding human population remains an enormous one.

References

1. Reay, D. S. et al. Global agriculture and nitrous oxide emissions. *Nature Climate Change* 2, 410–416, doi:10.1038/nclimate1458 (2012).
2. Solomon, S. *Climate change 2007 – the physical science basis: Working group I contribution to the fourth assessment report of the IPCC*. Vol. 4 (Cambridge University Press, 2007).
3. Smil, V. *Enriching the earth: Fritz Haber, Carl Bosch, and the transformation of world food production* (MIT Press, 2001).
4. Erisman, J. W., Sutton, M. A., Galloway, J., Klimont, Z. & Winiwarter, W. How a century of ammonia synthesis changed the world. *Nature Geoscience* 1, 636–639, doi:10.1038/ngeo325 (2008).
5. Lawes, J. B., Gilbert, J. H. & Rothamsted Experimental Station. *Determinations of nitrogen in the soils of some of the experimental fields at Rothamsted: and the bearing of the results on the question of the sources of the nitrogen of our crops.* (Harrison and Sons, 1883).
6. Tabashnik, B. E. Responses of pest and non-pest Colias butterfly larvae to intraspecific variation in leaf nitrogen and water content. *Oecologia* 55, 389–394 (1982).
7. Huber, D. & Watson, R. Nitrogen form and plant disease. *Annual Review of Phytopathology* 12, 139–165 (1974).
8. Batjes, N. H. Total carbon and nitrogen in the soils of the world. *European Journal of Soil Science* 47, 151–163 (1996).
9. Galloway, J. N. The global nitrogen cycle: past, present and future. *Science in China. Series C, Life Sciences/Chinese Academy of Sciences* 48 Spec No, 669–677 (2005).

10. Saikh, H., Varadachari, C. & Ghosh, K. Changes in carbon, nitrogen and phosphorus levels due to deforestation and cultivation: a case study in Simlipal National Park, India. *Plant and Soil* **198**, 137–145 (1998).
11. Reay, D. S., Dentener, F., Smith, P., Grace, J. & Feely, R. A. Global nitrogen deposition and carbon sinks. *Nature Geoscience* **1**, 430–437, doi:10.1038/ngeo230 (2008).
12. Sutton, M., Asman, W. & Schjoerring, J. Dry deposition of reduced nitrogen. *Tellus B* **46**, 255–273 (1994).
13. Hanson, P. J. & Lindberg, S. E. Dry deposition of reactive nitrogen compounds: a review of leaf, canopy and non-foliar measurements. *Atmospheric Environment. Part A. General Topics* **25**, 1615–1634 (1991).
14. Lovett, G. M. & Lindberg, S. E. Atmospheric deposition and canopy interactions of nitrogen in forests. *Canadian Journal of Forest Research* **23**, 1603–1616 (1993).
15. Tuck, A. Production of nitrogen oxides by lightning discharges. *Quarterly Journal of the Royal Meteorological Society* **102**, 749–755 (1976).
16. Thornton, P. E., Lamarque, J. F., Rosenbloom, N. A. & Mahowald, N. M. Influence of carbon-nitrogen cycle coupling on land model response to CO2 fertilization and climate variability. *Global Biogeochemical Cycles* **21** (2007).
17. Churkina, G., Trusilova, K., Vetter, M. & Dentener, F. Contributions of nitrogen deposition and forest regrowth to terrestrial carbon uptake. *Carbon Balance and Management* **2** (2007).
18. Vitousek, P. M. & Howarth, R. W. Nitrogen limitation on land and in the sea: how can it occur? *Biogeochemistry* **13**, 87–115 (1991).
19. Turner, D. P., Koerper, G. J., Harmon, M. E. & Lee, J. J. A carbon budget for forests of the conterminous United States. *Ecological Applications* **5**, 421–436 (1995).
20. Nadelhoffer, K. J. et al. Nitrogen deposition makes a minor contribution to carbon sequestration in temperate forests. *Nature* **398**, 145–148 (1999).
21. Magnani, F. et al. The human footprint in the carbon cycle of temperate and boreal forests. *Nature* **447**, 849–851 (2007).
22. Dentener, F. et al. Nitrogen and sulfur deposition on regional and global scales: a multimodel evaluation. *Global Biogeochemical Cycles* **20**, GB4003 (2006).
23. De Vries, W., Reinds, G. J., Gundersen, P. & Sterba, H. The impact of nitrogen deposition on carbon sequestration in European forests and forest soils. *Global Change Biology* **12**, 1151–1173 (2006).
24. Huttunen, J. T. et al. Fluxes of methane, carbon dioxide and nitrous oxide in boreal lakes and potential anthropogenic effects on the aquatic greenhouse gas emissions. *Chemosphere* **52**, 609–621 (2003).
25. Achtnich, C., Bak, F. & Conrad, R. Competition for electron donors among nitrate reducers, ferric iron reducers, sulfate reducers, and methanogens in anoxic paddy soil. *Biology and Fertility of Soils* **19**, 65–72 (1995).
26. Klüber, H. D. & Conrad, R. Effects of nitrate, nitrite, NO and N2O on methanogenesis and other redox processes in anoxic rice field soil. *FEMS Microbiology Ecology* **25**, 301–318 (1998).
27. Reay, D. S. & Nedwell, D. B. Methane oxidation in temperate soils: effects of inorganic N. *Soil Biology & Biochemistry* **36**, 2059–2065, doi:10.1016/j.soilbio.2004.06.002 (2004).

28. Reay, D. S., Nedwell, D. B., McNamara, N. & Ineson, P. Effect of tree species on methane and ammonium oxidation capacity in forest soils. *Soil Biology & Biochemistry* **37**, 719–730, doi:10.1016/j.soilbio.2004.10.004 (2005).

29. Mosier, A., Schimel, D., Valentine, D., Bronson, K. & Parton, W. Methane and nitrous oxide fluxes in native, fertilized and cultivated grasslands. *Nature* **350**, 330–332 (1991).

30. Bodelier, P. L. & Laanbroek, H. J. Nitrogen as a regulatory factor of methane oxidation in soils and sediments. *FEMS Microbiology Ecology* **47**, 265–277 (2004).

31. Eckard, R. & Cullen, B. Impacts of future climate scenarios on nitrous oxide emissions from pasture based dairy systems in south eastern Australia. *Animal Feed Science and Technology* **166**, 736–748 (2011).

32. Abdalla, M. et al. Testing DayCent and DNDC model simulations of N_2O fluxes and assessing the impacts of climate change on the gas flux and biomass production from a humid pasture. *Atmospheric Environment* **44**, 2961–2970 (2010).

33. Kamp, T., Steindl, H., Hantschel, R., Beese, F. & Munch, J.-C. Nitrous oxide emissions from a fallow and wheat field as affected by increased soil temperatures. *Biology and Fertility of Soils* **27**, 307–314 (1998).

34. Cantarel, A. A., Bloor, J. M., Deltroy, N. & Soussana, J.-F. Effects of climate change drivers on nitrous oxide fluxes in an upland temperate grassland. *Ecosystems* **14**, 223–233 (2011).

35. Parry, M. L., Rosenzweig, C., Iglesias, A., Livermore, M. & Fischer, G. Effects of climate change on global food production under SRES emissions and socio-economic scenarios. *Global Environmental Change* **14**, 53–67 (2004).

36. Davidson, E. A. Representative concentration pathways and mitigation scenarios for nitrous oxide. *Environmental Research Letters* **7**, 024005 (2012).

37. Olesen, J. E. et al. Uncertainties in projected impacts of climate change on European agriculture and terrestrial ecosystems based on scenarios from regional climate models. *Climatic Change* **81**, 123–143 (2007).

38. Mkhabela, M., Gordon, R., Burton, D., Smith, E. & Madani, A. The impact of management practices and meteorological conditions on ammonia and nitrous oxide emissions following application of hog slurry to forage grass in Nova Scotia. *Agriculture, Ecosystems & Environment* **130**, 41–49 (2009).

39. Sommer, S. G. et al. Processes controlling ammonia emission from livestock slurry in the field. *European Journal of Agronomy* **19**, 465–486 (2003).

40. Suddick, E. & Davidson, E. The role of nitrogen in climate change and the impacts of nitrogen-climate interactions on terrestrial and aquatic ecosystems, agriculture, and human health in the United States: a technical report submitted to the US National Climate Assessment. *North American Nitrogen Center of the International Nitrogen Initiative (NANC-INI), Woods Hole Research Center* **149**, 208 (2012).

41. Butterbach-Bahl, K. et al. Nitrogen as a threat to the European greenhouse balance. In *The European Nitrogen Assessment: Sources, Effects and Policy Perspectives*, edited by M. A. Sutton et al., Chap. 19, 434–462 (Cambridge University Press, UK, 2011).

42. Butterbach-Bahl, K. et al. Nitrogen processes in terrestrial ecosystems. In *The European Nitrogen Assessment: Sources, Effects and Policy Perspectives*, edited by M. A. Sutton et al., Chap. 6, 99–125 (Cambridge University Press, UK, 2011).

43. Bobbink, R., Hornung, M. & Roelofs, J. G. The effects of air-borne nitrogen pollutants on species diversity in natural and semi-natural European vegetation. *Journal of Ecology* 86, 717–738 (1998).
44. Bobbink, R. et al. Global assessment of nitrogen deposition effects on terrestrial plant diversity: a synthesis. *Ecological Applications: A Publication of the Ecological Society of America* 20, 30–59 (2010).
45. Throop, H. L. & Lerdau, M. T. Effects of nitrogen deposition on insect herbivory: implications for community and ecosystem processes. *Ecosystems* 7, 109–133 (2004).
46. Bale, J. S. et al. Herbivory in global climate change research: direct effects of rising temperature on insect herbivores. *Global Change Biology* 8, 1–16 (2002).
47. Thomas, C. D. et al. Extinction risk from climate change. *Nature* 427, 145–148 (2004).
48. Sala, O. E. et al. Global biodiversity scenarios for the year 2100. *Science* 287, 1770–1774 (2000).
49. Phoenix, G. K. et al. Atmospheric nitrogen deposition in world biodiversity hotspots: the need for a greater global perspective in assessing N deposition impacts. *Global Change Biology* 12, 470–476 (2006).
50. Keuken, M., Bakker, F., Möls, J., Broersen, B. & Slanina, J. Atmospheric deposition and conversion of ammonium to nitric acid on a historic building: a pilot study. *International Journal of Environmental Analytical Chemistry* 38, 47–62 (1990).
51. Vitousek, P. M. et al. Human alteration of the global nitrogen cycle: sources and consequences. *Ecological Applications* 7, 737–750 (1997).
52. Fangmeier, A., Hadwiger-Fangmeier, A., Van der Eerden, L. & Jäger, H.-J. Effects of atmospheric ammonia on vegetation – a review. *Environmental Pollution* 86, 43–82 (1994).

7
Freshwater Nitrogen and Climate Change

Snoozing and farting on the beaches that encircle the forbidding island of South Georgia are some of the most malodorous animals alive. It is here, and on similar islands and sheltered bays dotted around Antarctica, that southern elephant seals come together in huge numbers[1]. With bellies full of the squid that abound in the rich ocean waters, they haul themselves ashore to mate, raise pups and sunbathe.

In the 1950s and 1960s, hunters used these slow-moving mounds of blubber as a ready alternative to the devastated stocks of great whales, the few accessible inlets of South Georgia still bearing testament to this bloody thirst for animal oil. Today the rusting harpoon sheds and blubber-boiling houses are home to gangs of female elephant seals, each vying for the sunniest wind-free positions. Amongst the heavy ropes and boat carcasses, they lie together like packs of overstuffed sausages, emitting a smell that is nothing short of eye-watering. South Georgia is home to more than half of the global elephant seal population – including some 100,000 shed-loving females – and they, along with the many penguins and fur seals that also enjoy these sheltered bays, bring with them waves of enriching nitrogen from the ocean[2]. Countless stomach loads of digested squid and shrimp end up spread over the beaches and the boggy wallowing areas that lie behind them.

These areas become increasingly fertile pockets in which algae and bacteria can feast on the bountiful flush of new nutrients, resulting in 'eutrophication'[3]. The further the lumbering seals move inland to vent their rotund stomachs, the more the once-virginal meltwater streams and pools become clogged with a bubbling mix of microbes. Greener ponds and even more foul-smelling seal wallows in and around Antarctica may be rather remote impacts of such creeping enrichment, but across much of the world the massive amounts of nitrogen flowing into streams, rivers and lakes are posing an increasing threat to all life, including humankind.

This freshwater eutrophication problem has now gone global. It is being seen in thousands of lakes in Africa and South America and in around half of all the lakes in Asia, Europe and North America[4]. The Millennium Ecosystem Assessment[5] report identified the growth in reactive nitrogen emissions – from 20 million tonnes a year in the early 1900s to 150 million tonnes a year today – as one of the greatest threats to the aquatic environment. The inefficient use of fertiliser and manure nitrogen, coupled with the ploughing up of established grasslands and an intensifying rain of reactive nitrogen deposition from the atmosphere, has contributed to the strengthening flush of reactive nitrogen into many freshwater ecosystems and to serious concerns over the resultant negative environmental, climatic and water quality impacts.

The starting point for much of the reactive nitrogen that now enters freshwater systems around the world is agriculture (Chapter 6) and the leaching and run-off of nitrogen-rich waters into groundwaters and drainage systems, called 'diffuse' nitrogen pollution[6]. Other diffuse sources include the reactive nitrogen released due to soil erosion and disturbance, and that introduced directly to rivers, lakes and reservoirs by deposition from the atmosphere (Chapter 5).

Their diffuse nature makes these sources difficult to monitor and manage, with large-scale control of land use and nitrogen applications often being required to dent the large cumulative flow of the myriad small sources of nitrogen pollution across a river catchment[7]. More concentrated sources of reactive nitrogen to freshwater systems are called 'point sources', and these include sewage treatment works and industrial sites where release of effluent into rivers can represent a major nitrogen input at a single location[8]. In much of the world untreated sewage from septic tanks and urban drainage can also lead to significant reactive nitrogen pollution of surface waters.

As human activities now dominate the global nitrogen cycle, they are also the primary driver of reactive nitrogen fluxes in aquatic systems. However, natural sources can also be important in some locations. Blooms of nitrogen-fixing algae in ponds, lakes and rice paddies[9,10] may provide large, and sometimes useful, reactive nitrogen inputs, while lightning-derived nitrate deposition[11] and flushes of nitrogen in snowmelt or upland storm water run-off[12] can represent further additions to human-induced fluxes (Figure 7.1).

One important pathway for freshwater nitrogen flows that can take both human-induced and natural forms is via fish. The global fish

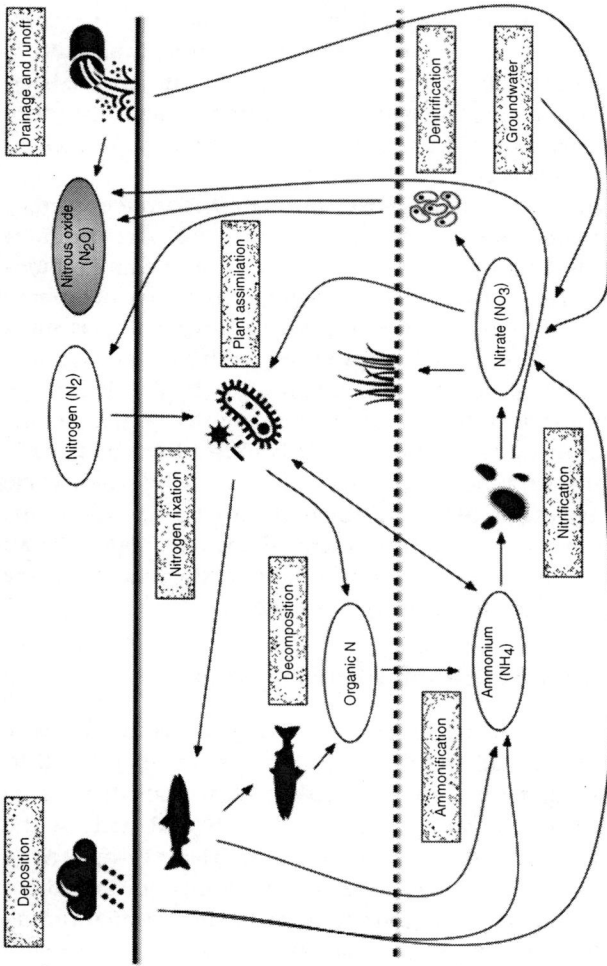

Figure 7.1 The freshwater nitrogen cycle

The solid line represents the water surface and the dotted line the stream, river or lake bed. The boxes show the key processes by which nitrogen is cycled and the ovals show the changing form of nitrogen as it undergoes these processes. Denitrification is a major process for converting the nitrate dissolved in freshwaters to inert dinitrogen gas (N_2). Anaerobic ammonium oxidation (called 'anammox') is also very important in freshwater environments (especially in sediments), with the anammox bacteria able to convert the ammonium and nitrite directly to dinitrogen.

Source: Dave Reay

population represents a living and dynamic stock of reactive nitrogen due to the abundant proteins and amino acids in fish tissues[13]. The natural migration of fish species such as salmon from seas to rivers can therefore introduce large amounts of nitrogen to freshwater systems. In some areas the sheer volume of fish that return from the sea to spawn and then die in rivers and streams makes them a dominant source of reactive nitrogen in the local ecosystem[14]. Similarly, freshwater aquaculture can directly introduce very large amounts of nitrogen into ponds, lakes and, to a lesser extent, rivers. Globally, aquaculture feed and fish waste is now estimated to add more than one million tonnes of nitrogen to freshwater systems every year[15].

Reactive nitrogen can be transported through freshwater systems either dissolved in the water or bound up in particles. The common dissolved forms include inorganic nitrogen such as nitrate and ammonium, with the nitrogen in particles often being in the form of organic nitrogen[16]. In surface waters it is the dissolved nitrogen that is most readily available for plants and microbes to use, with flushes of nitrate and ammonium allowing aquatic plants and algae to flourish. Once converted into biomass the nitrogen can then be recycled, as the organic nitrogen in plant leaves and dead algal cells is mineralised by aquatic microbes to form nitrate and ammonium, and so start the whole cycle once again[17]. This cycling of nitrogen through freshwater systems means that one nitrogen atom may be used many times and appear in all parts of the aquatic nitrogen cycle from mountain streams to estuary mouths – commonly called the 'nitrogen cascade'[18].

Climate change and freshwater nitrogen cycling

Arguably the most overt impacts of climate change in the 21st century will be via changes in rainfall and the freshwater systems that it feeds[19]. Increasing global temperatures are expected to lead to an increase in annual precipitation in the wetter regions of the world such as the temperate and tropical latitudes and a decrease in drier regions such as the sub-tropics. Within these broad-scale changes a shift towards more extreme events is expected[20], alongside changes in the seasonal timing and volume of meltwater run-off from glaciers and high-altitude snow[21]. These alterations in the water cycle have major implications for how much reactive nitrogen enters freshwater systems as well as its processing and its impacts[22–24]. Where precipitation and meltwater flows decline, there is a greater risk of reactive nitrogen concentrations increasing to a level where they have negative effects on water quality and freshwater biodiversity[25,26]. This poses a serious challenge for industries that must meet set standards for nitrogen concentrations in water, such as the sewage and water treatment sectors. Here, low water flows in the future may mean that regulations are breached and large costs incurred. Drought and low water flow events can also lead to changes in the sourcing of water for human use, with a shift towards more reliance on high-nitrate groundwaters[27,28] and potentially increased human health risks.

High precipitation and water flow events pose major challenges for freshwater nitrogen management too, as larger amounts of reactive nitrogen are often flushed into and through the system. This can result in a much larger amount of reactive nitrogen reaching estuarine and coastal areas instead of being processed and removed by denitrification en route[29].

Rising temperatures are expected to increase the rate of nitrogen processing in many freshwater systems, warmer water and sediments allowing more rapid cycling of nitrogen by plants and microbes. However, in static or very slow-moving water bodies, such as lakes, rising temperatures may increase stratification and so deprive algae in surface waters of the nitrogen supplied by deeper waters and sediments[29]. This may result in a reduction in plant growth near the surface and an increase in low-oxygen or anoxic conditions in deeper waters. In polar regions, more thawing of ice and permafrost can also mean additional water flow and release of stored reactive nitrogen at the top of catchments and therefore more eutrophication issues downstream[30] (Figure 7.2). Another potentially important interaction of climate change and freshwater nitrogen is in the enhancement of flood risk. As nutrient supplies to aquatic plants increase, plant growth rates may rise to a level where drainage streams and channels become choked with plants and debris[31], thereby leading to much more severe flooding during storm flow events.

Eutrophication and biodiversity

As climate change intensifies, the increase in freshwater plant growth associated with greater reactive nitrogen inputs to freshwaters (i.e. eutrophication) may have serious implications for aquatic biodiversity. Intense algal blooms can smother other plant species[22], while those aquatic plants adapted for low-nutrient (called 'oligotrophic') conditions may be outcompeted by more nitrogen-loving species[32]. At the same time, climate change will increase water temperatures and alter the timing and magnitude of freshwater flows in many regions. These impacts may eliminate some species and force others to migrate to new areas. Together, eutrophication and climate changes can have knock-on

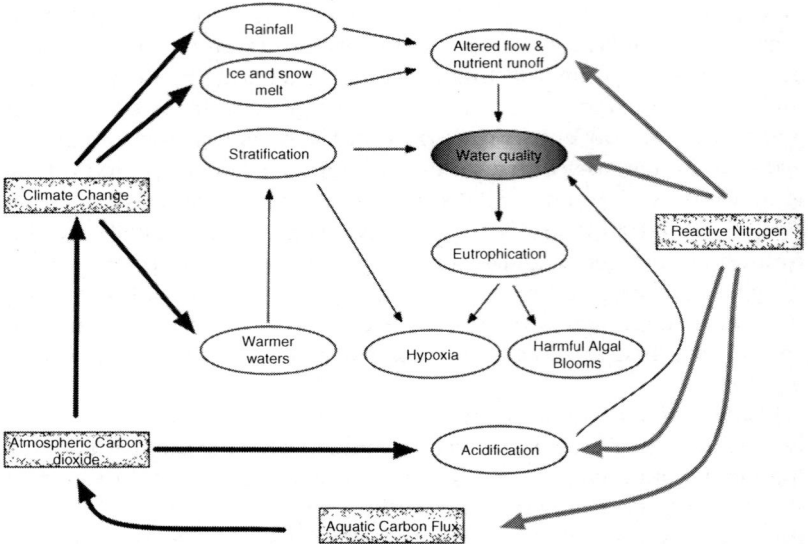

Figure 7.2   Key interactions of nitrogen and climate change in aquatic environments

The major impacts of climate change and increasing carbon dioxide concentrations (large black arrows) and the major impacts of nitrogen (large grey arrows) come together directly in freshwater systems through effects such as poor water quality (from more nitrogen inputs and increased run-off). High nitrate concentrations are a particular concern for drinking water supplies.

Source: Dave Reay

effects for the entire freshwater ecosystem food chain[33], radically altering the wider community of plants, invertebrates and fish.

Anoxic waters resulting from enhanced plant growth, higher decomposition rates and altered water circulation can also have negative impacts for freshwater biodiversity, with plants and animals that live on or in lake beds (called 'benthic' species') being at particular risk[34,35]. The algal blooms that result from eutrophication may be comprised of algae that produce toxins dangerous to animal and plant life alike and are called 'harmful algal blooms'. In large rivers, lakes and reservoirs such harmful algal blooms pose a significant threat to fisheries, biodiversity, recreational use and water quality[36].

The eutrophication of freshwater systems (primarily via increasing nitrogen and phosphorous inputs) is already a major concern in many regions of the world. As climate change intensifies in the coming decades and the hydrological cycle becomes more and more disrupted, the negative consequences of increasing reactive nitrogen inputs to freshwater systems risk being exacerbated even further[37].

Freshwater and nitrous oxide

Leaching and run-off of nitrogen, usually as nitrate from agriculture, represents a significant global source of the powerful greenhouse gas nitrous oxide. Nitrogen processing in streams, rivers and lakes leads, through nitrification and denitrification, to the annual formation and emission of around one million tonnes of nitrogen in the form of nitrous oxide[38]. When high rates of nitrous oxide formation in soils coincide with very wet or saturated soils, substantial amounts may also enter freshwater systems via field drainage systems[39–41]. It is estimated that around 30 per cent of the nitrogen applied to agricultural soils is subsequently lost in water run-off and soil leachate[42], but the precise amounts of nitrous emissions then produced and the factors that control such emissions from these indirect routes are still rather uncertain[43].

Freshwater aquaculture, and its related nitrogen inputs of more than one million tonnes a year[15], also represents a significant source of freshwater nitrous oxide emissions. Intensive fish and shellfish rearing in ponds, lakes and reservoirs can provide ideal conditions for nitrous oxide production, with a recent estimate putting global emissions from aquaculture at around 120,000 tonnes of nitrogen (0.12 Tg N) per year[44]. The rapid growth of this sector makes it very likely that the proportion of freshwater nitrous oxide emissions for which it is responsible will also continue to rise.

There are numerous stages at which nitrous oxide can be both formed and emitted to the atmosphere in freshwater ecosystems. The inorganic nitrogen load in drainage waters, arising from leaching and run-off, may not encounter conditions suitable for sediment nitrification or denitrification until many kilometres downstream from the point source[45]. Consequently, the total amount of nitrous oxide emission attributable to a given amount of nitrogen pollution in the water may be spread over many kilometres, making its quantification very difficult (Box 7.1).

The outgassing of dissolved nitrous oxide coming from soil leachate tends to be most pronounced in agricultural drainage streams, while production of nitrous oxide from nitrate carried downstream in the drainage waters tends to become more important in rivers and estuaries[16,54]. Globally, freshwater systems are thought to be responsible for around 20 per cent of all denitrification on the planet, making the millions of streams, lakes and rivers a key component of the global nitrogen cycle[38].

Box 7.1

The 'leaky pipe'

As the various forms of nitrogen are cycled through aquatic systems, some can be temporarily or permanently removed at each step. In environments such as flooded rice paddies or managed waterways, some of the nitrogen that is incorporated into the plants and sediments is removed during harvesting or vegetation clearance[46]. In large lakes and reservoirs the nitrogen bound up in dead biomass and other particles may also become locked away in sediments for many years or decades[47,48]. The most important route for nitrogen removal from many freshwater systems is to the atmosphere and, just as in terrestrial ecosystems, nitrification and denitrification are the key pathways[49]. Occurring either in the water column or in sediments, these microbial processes are able to convert inorganic nitrogen to nitrous oxide and inert nitrogen gas, which can then escape to the atmosphere[50]. The climate change impacts of and interactions with the freshwater nitrogen cascade include a range of indirect effects, such as increasing snow and ice melt rates changing how much nitrogen is flushed into meltwater lakes and rivers[51]. However, it is the 'leaky pipe' of nitrogen processing, and specifically the emissions of nitrous oxide[52,53], that represents the most direct involvement of freshwater nitrogen in global climate change[38].

Climate change is expected to alter the timing of inputs, volumes and concentrations of reactive nitrogen in these freshwater ecosystems[22], thereby affecting both nitrous oxide production and emissions. Temperature increases will also alter microbial nitrogen processing rates, stratification (especially in lakes and reservoirs) and oxygen availability in freshwater systems[55,56] – all of which can affect nitrous oxide production and emissions. Whether the net global impact of climate change will be an increase or decrease in freshwater-derived nitrous oxide emissions remains uncertain.

Nitrogen and the freshwater carbon fluxes

Just as increased reactive nitrogen inputs can accelerate plant and algal growth rates in freshwater systems, they also have the potential to increase the amount of carbon dioxide that is taken up via primary production and the amount of carbon stored in river, lake and reservoir sediments. In waters that are already rich in nitrogen this enhancement of plant growth and carbon uptake is limited by the supply of other nutrients, such as phosphorus[57]. The true size of the global freshwater carbon sink is uncertain, but the world's lakes alone are estimated to be a net sink for carbon of 20–70 million tonnes (0.02–0.07 Pg C) each year[58,59]. Where increased reactive nitrogen inputs do result in large increases in plant growth[60] and carbon storage in freshwater sediments[61], this process may be a significant help in efforts to stabilise carbon dioxide concentrations in the atmosphere. However, any reduced climate forcing achieved by a larger freshwater carbon sink must be set against the likely increase in freshwater nitrous oxide emissions[62]. Climate change itself will also affect the global freshwater carbon sink, with increasing water temperatures leading to accelerated rates of carbon cycling and changing water flows, altering nitrogen and carbon inputs[63].

Methane and freshwater nitrogen

Though nitrous oxide is the primary greenhouse gas emitted as a result of increasing nitrogen loading of freshwater ecosystems, fluxes of methane – another powerful greenhouse gas – from these systems can be substantial, and in part determined by nitrogen[64,65]. The bulk of methane is produced by microbes in the low-oxygen or anoxic conditions common to aquatic sediments. These microbes (called 'methanogens') use simple carbon sources, such as carbon dioxide and acetate, to form methane in the absence of oxygen[66]. An increase in nitrogen inputs can enhance methane production by boosting plant growth and thereby increase the supply of carbon to the methanogens[67,68]. An increase in the amount of biomass decomposing in the water and sediments may also lead to lower oxygen concentrations[69], resulting in conditions even more conducive to methane production. Again, climate change impacts may exacerbate these issues by increasing sediment and water temperatures (so accelerating methanogenesis) and by increasing the carbon and nitrogen inputs to freshwater systems via leaching and run-off.

Freshwater nitrogen and human health

There appear to be significant human health implications of nitrogen enrichment arising from its enhancement of freshwater-mediated pests and diseases, such as mosquitoes and the West Nile virus they transmit[70,71]. Similarly, the promotion of harmful algal blooms by high nitrogen availability risks increasing the exposure of humans to algal toxins in bathing water and drinking water and in the tissues of freshwater fish and shellfish. The direct linkage between freshwater nitrogen and human health impacts in these areas remains difficult to establish, as other nutrients (especially phosphorus), climate change and human behaviour are all important in determining the eventual impact. Much more direct, and certainly better studied, are the human health effects of nitrogen consumption itself.

Nitrogen consumption and human health

Nitrate contamination of drinking water is now an issue in many areas of the world[72]. Water companies spend a great deal of time and money blending water to dilute the levels of nitrate down to what is perceived as a safe level[73], yet many millions of people get their drinking water from private wells where concentrations of nitrate may be many times the recommended level. In the US more than one in five of the water wells in agricultural areas exceed the national limits, with many surface and groundwater sources in Europe having the same problem[74]. It is the excessive use of nitrogen fertilisers that causes most of this contamination, but nitrogen-rich sewage seeping into water supplies can also be a problem, especially in the towns and cities of the developing world. Whether from agriculture, sewage or some other source, the contamination of drinking water supplies with reactive nitrogen risks being exacerbated by the impacts of climate change, with more intense rainfall events increasing the losses of nitrogen (especially in the form of nitrate) to surface waters and groundwaters.

What exactly represents a dangerous concentration of nitrate in drinking water and food continues to kick up storms of scientific argument[75–77]. Nitrate and other forms of reactive nitrogen have a long history of use as medicines (Box 7.2), with some arguing that current limits on nitrate intake risk missing out on beneficial effects or adding unnecessary water-processing costs[78]. Others point to the links between high nitrate concentrations in drinking water and food and the occurrence of serious medical conditions such as blue baby syndrome[79] and some cancers[80].

It was in the 19th century that nitrate's crown as a universal panacea for medical ills began to slip. Other, more effective, treatments began to emerge, and the use of nitrate increasingly became limited to the treatment of inflammation and as a diuretic. One problem was the incredibly high doses that were usually required to achieve the desired effect. One doctor is recorded as having prescribed his rheumatic patients with such large doses that they had to be given in highly diluted portions to avoid corrosion of their digestive tracts. This saturation treatment was apparently very successful, though the patients had to endure bouts of nausea and diarrhoea along the way.

Box 7.2

Nitrogen and human health

Nitrogen and its role in human health has a long and rather chequered history. Alongside their discovery of the explosive properties of the naturally occurring nitrate (called 'nitre') found in the crusty desert deposits of China's northwestern provinces, the Chinese were also the first to record its use as a medicine. It was an ingredient in some of the Daoist elixirs of immortality, and as early as the 8th century it was apparently being prescribed as a treatment for angina attacks[81].

A few centuries later it was cropping up as a medicine in the Arab world, and by the 17th century it was gaining acclaim in Europe for its ability to 'excite the urine, whet the appetite and calm ardours', amongst other attributes[82]. What 'excited urine' actually involved and whether this was a desirable effect is unclear. The number of maladies that nitrate was prescribed for blossomed still further in the 18th century, when it was used to treat everything from bubonic plague to 'hysterical vapours and uterine fury'.

That nitrate had gained such a cure-all reputation is somewhat surprising, given that it is a relatively inert substance and should not interact much during its journey through the human body. The secret of its success – and ultimately its medical downfall – was what it can be turned into during this journey.

In a healthy adult human the nitrate consumed in food and drink or as a medicine is quickly absorbed from the gut into the bloodstream, where it joins still more nitrate that is naturally produced within the body. Most ends up expelled in urine, but some also finds its way into saliva and sweat. The concentrations in saliva can be very high, and the colonies of bacteria that live in all human mouths are able to convert these flushes of nitrate into the much more active form called nitrite – the same form that gives cured meat its distinctive pink colour. It is this nitrite that is at the heart of both the medical benefits of nitrate as a medicine and its negative side effects. Under the acid conditions of the human mouth and stomach the nitrite can be further transformed into various forms – in particular to nitric oxide[78] – that kill off harmful bacteria and help to increase blood flow. Such antibacterial properties of nitrite-rich saliva may also explain, in part, why animals lick wounds. Helping to prevent tooth decay and stomach upsets and to free up blood flow are among many potential benefits of this conversion process. On the skin too, nitrite and its products

(Continued)

Box 7.2 (Continued)

may help to wipe out any unwanted microbial invaders before they can take hold[83]. The latter seems to have been a well-publicised property, and as early as the 16th century the taking of nitrate was claimed to 'restore the skinne and complexion to the native bewtie'[81].

The medical use of nitrates was therefore generally down to either contamination of the original medicine with nitrite or the conversion of the prescribed nitrate to nitrite after it was taken. The angina sufferer in 8th-century China was given specific instructions to place the nitrate-rich medicine under his tongue and then swallow the saliva – presumably to take advantage of the busy conversion to nitrite by bacteria in the mouth.

The very rapid action of nitrite allowed sufferers the window they needed to rest and recover from an attack. The most commonly used form by the 19th century was amyl nitrite. However, the volatile nature of amyl nitrite made it difficult to use, and its effects were very short-lived. Nitroglycerine may seem an odd choice as a more stable modern substitute, but by the end of the 19th century this highly explosive formulation discovered by Ascanio Sobrero, and subsequently tamed by Alfred Nobel, was fast becoming the angina treatment of choice[84]. Sobrero was the first to report some non-explosive effects of his discovery, citing the intense headache that even small amounts induced when placed under the tongue. Homeopathic medicine then led the way in testing these medical impacts and potential benefits, and by the time that Nobel himself reported to his doctor with angina pains in 1895 it was nitroglycerine that was prescribed. He refused it.

Through the 20th century nitroglycerine became increasingly dominant as an angina treatment, though exactly how it worked remained a mystery. One clue to this mechanism lay in the increasing frequency of the aptly named 'Monday disease' suffered by workers exposed to nitroglycerine in the flourishing explosives industry[85]. During the working week these workers received a continual dose of nitroglycerine that dilated their blood vessels, just as it does for angina sufferers. At the weekend, with the dustings of nitroglycerine cut off, the blood vessels contracted and reduced the blood supply to the heart, bringing on bouts of angina. The mystery of its action was finally solved by Robert Furchott, Louis Ignarro and Ferid Murad, who established that it was the nitric oxide produced by nitroglycerine that served to relax smooth muscle cells and so lower blood pressure and wipe away the angina[86]. Fittingly, they each received the Nobel Prize for their discovery. Today millions of people rely on so-called organic 'nitrates' for rapid relief of their angina, nitroglycerine still being the most common form.

Because of the belief that nitrate could reduce inflammation, it was also often used as a handy treatment for the burning discomfort of gonorrhoea. Afflicted soldiers wanting to avoid difficult questions and a pricey physician's bill were known to dissolve the nitrate-rich contents of a rifle cartridge in a glass of water for self-administered relief. Indeed a widespread rumour of barracks life was that nitrate was deliberately added to soldiers' food to reduce their libido and so help prevent the spread of sexually transmitted diseases.

The medical use of such inorganic forms of nitrate – the naturally occurring types like nitre and the manufactured products like ammonium nitrate – dwindled still further in the 20th century. Other than some trials on nitrate as a diuretic in the US in the 1930s, they largely disappeared from the pharmacies and medicine chests of the world as worries about their side effects intensified. Chief among these, and one that is still keeping nitrate on the medical back-burner, is the shockingly overt and occasionally deadly condition known as methemoglobinemia or blue baby syndrome.

Blue baby syndrome

Nitrite, that product of nitrate conversion in the human body that can be so beneficial to health, is not always so good. Where the bubbling acid-bath conditions of the stomach have been calmed by indigestion remedies, or the intestines have been colonised by more than the usual array of nitrate-transforming bacteria, the amounts of nitrite that begin to swill around the body can climb higher and higher. In the blood stream it can then bind to the haemoglobin in red blood cells and prevent them from carrying their usual payload of oxygen around the body. Normally, the amounts of haemoglobin-clogging nitrite are small and plenty of oxygen is still able to get where it needs to be, but if levels rise high enough the ability of the blood to carry oxygen can start to plummet[87].

The first sign that tissues are being starved of oxygen is a blue tinge to the skin and lips, then come feelings of anxiety, headaches and a racing heart. If left untreated, the symptoms can progress through fatigue and confusion into seizures, coma and ultimately death. Going by the tongue-tangling medical name of methemoglobinemia, this blue baby syndrome is, as its name suggests, most common in infants. Young children are hit by the double whammy of high fluid intakes, with the inherent risk that these fluids will contain large amounts of nitrate or nitrite, and the limited ability to convert the nitrite-bound haemoglobin back into the oxygen-carrying form.

A link between nitrate-rich well water and the development of blue baby syndrome was first made in the US in 1945[88]. Since then there have been around 3,000 cases recorded worldwide, and most of these have been associated with private water wells containing high levels of nitrate. As a result, a 'safe' level for nitrate in drinking water (less than 50 parts per million) has been applied in most countries. Yet over the years factors other than simply how much nitrate is consumed have been suggested as more important causes of blue baby syndrome[89]. These include contamination of wells and feeding bottles by bacteria and the development of the illness in babies already suffering from diarrhoea or respiratory disease, with any extra nitrate intake serving only to exacerbate existing problems.

High levels of nitrate in drinking water may therefore bump up the severity of blue baby syndrome cases, if not lead to their direct occurrence. Avoiding private well water when making up formula milk is an obvious way of reducing this risk, but several cases show that even where the water quality is fine, babies can receive large doses of nitrate in the outwardly healthy form of vegetables[90]. In the average Western diet, more than half of all the nitrate consumed will be contained in vegetables, rather than drinks. Some, like spinach, lettuce and carrots, can be very nitrate-rich – in Europe there are now limits on the amount of nitrate in lettuce and spinach, as well as in water. In two recent cases of blue baby syndrome it was a bowel-loosening meal of courgette soup that apparently delivered the life-threatening dose of nitrate[91]. Both babies, one a month-old boy, the other a two-month-old girl, were initially breast-fed and had developed normally. When their mothers switched over to formula milk they played safe and used bottled water, but when the babies then became constipated their doctor recommended they use courgette soup to reconstitute the formula milk. Within a few weeks both infants began to go downhill, their skin becoming mottled and cold to the touch, their behaviour getting increasingly irritable. On admission to hospital the levels of oxygen in their blood were found to be worrying low, and methemoglobinemia was quickly diagnosed. Both babies were treated with methylene blue – the rather ironic treatment for such cases that works by converting the nitrite-bound haemoglobin back into its oxygen-carrying form – and each fully recovered within less than a day. Though most often occurring in children younger than 12 months, it is not always babies that turn blue (Box 7.3).

As mentioned earlier, blue baby syndrome and the ongoing resistance to more widespread use of nitrate as a medical treatment that it engenders incurs the risk that opportunities to develop new medicines may be missed[75,78,82]. The beneficial effects of nitrate and nitrite as well as their products on the human body have the potential to be deliberately

Box 7.3

Blue baby syndrome in adults

In a recent case in Italy[92], it was a 23-year-old woman who was rushed into the emergency department of the local hospital by her father with signs that blue baby syndrome was taking hold. The woman was confused and disoriented, her skin was the colour of slate, and her tongue and lips were almost black. She was complaining of stomach pain and her heart was racing at 140 beats per minute. Initial tests drew a blank. Ultrasound testing showed her stomach to be normal and there was no evidence that she had taken any drugs. Oxygen levels in her blood were low and failed to improve even with an oxygen mask, and her condition worsened. The distinctive chocolate-brown colour of the blood in her arteries suggested blue baby syndrome and, with the situation becoming increasingly grave, her doctors decided they could wait no longer and gave her a dose of methylene blue. It worked. Within minutes the colour began to return to her skin; she became less confused and was soon able to speak fluently. It was then that she was able to tell the doctors what had happened.

The evening before, she had eaten a meal of very salty rice for dinner. Waking at midnight with a raging thirst, she had gone to the kitchen to get a drink. There she had scraped chunks of ice from the inside of the freezer into a glass of tap water, had drunk deeply and gone back to bed. As the night wore on her stomach became sore, she was hit with bouts of diarrhoea and by the morning her lips were already turning blue. Though becoming increasingly ill, she still went into work and, once there, fainted twice before her father arrived and took her to hospital.

The key to her illness was tracked down to the chunks of ice scraped from the freezer for her glass of water. A few days before the woman had had a bad toothache and her dentist had given her an instant ice pack to help ease the pain. After using it she had put the pack into the freezer. The pack had burst, its contents freezing into the icy lumps that had then cooled her thirst-quenching midnight drink. The instant ice was made of ammonium nitrate. Such cases of blue baby syndrome in adults are very rare and, barring draughts of concentrated ammonium nitrate, the adult body is usually able to keep the levels of nitrite-hobbled haemoglobin to a minimum.

exploited to increase blood flow in target areas or to help wipe out bacterial infections. In years to come nitrate and nitrite may well begin to shed their predominantly negative associations, but blue baby syndrome is not the only downside to which they have been linked. As well as the pathogen-killing, artery-opening benefits of nitric oxide, nitrite in the body may lead to the generation of a group of gene-bending forms of nitrogen called 'N-nitroso compounds'[93]. These chemicals have been implicated in cases of stomach, bladder and colon cancer to non-Hodgkin lymphoma and to birth defects[87,94]. Though the direct link between high nitrate intake and an increased risk of such cancers and defects remains controversial, the debate over the human health costs and benefits of nitrate in our water and food looks set to run and run.

References

1. Boyd, I., Walker, T. & Poncet, J. Status of southern elephant seals at South Georgia. *Antarctic Science* 8, 237–244 (1996).
2. Whitehouse, M., Priddle, J., Brandon, M. & Swanson, C. A comparison of chlorophyll/nutrient dynamics at two survey sites near South Georgia, and the potential role of planktonic nitrogen recycled by land-based predators. *Limnology and Oceanography* 44, 1498–1508 (1999).
3. Lewis Smith, R. Destruction of Antarctic terrestrial ecosystems by a rapidly increasing fur seal population. *Biological Conservation* 45, 55–72 (1988).
4. Smith, V. H. Eutrophication of freshwater and coastal marine ecosystems a global problem. *Environmental Science and Pollution Research* 10, 126–139 (2003).
5. Assessment, M. E. *Ecosystems and human well-being.* Vol. 5 (Island Press, Washington, DC, 2005).
6. Reay, D. S., Smith, K. A. & Edwards, A. C. Nitrous oxide emission from agricultural drainage waters. *Global Change Biology* 9, 195–203, doi:10.1046/j.1365-2486.2003.00584.x (2003).
7. Brown Gaddis, E. J., Vladich, H. & Voinov, A. Participatory modeling and the dilemma of diffuse nitrogen management in a residential watershed. *Environmental Modelling & Software* 22, 619–629 (2007).
8. Puckett, L. J. Identifying the major sources of nutrient water pollution. *Environmental Science & Technology* 29, 408A–414A (1995).
9. De, P. The role of blue-green algae in nitrogen fixation in rice-fields. *Proceedings of the Royal Society of London. Series B, Biological Sciences* 127, 121–139 (1939).
10. Roger, P.-A. & Ladha, J. In *Biological nitrogen fixation for sustainable agriculture,* 41–55 (Springer, 1992).
11. Schumann, U. & Huntrieser, H. The global lightning-induced nitrogen oxides source. *Atmospheric Chemistry and Physics* 7, 3823–3907 (2007).
12. Brooks, P. D., Campbell, D. H., Tonnessen, K. A. & Heuer, K. Natural variability in N export from headwater catchments: snow cover controls on ecosystem N retention. *Hydrological Processes* 13, 2191–2201 (1999).

13. Deegan, L. A. Nutrient and energy transport between estuaries and coastal marine ecosystems by fish migration. *Canadian Journal of Fisheries and Aquatic Sciences* 50, 74–79 (1993).

14. Hocking, M. D. & Reimchen, T. E. Salmon-derived nitrogen in terrestrial invertebrates from coniferous forests of the Pacific Northwest. *BMC Ecology* 2, 4 (2002).

15. Bouwman, A. F. et al. Hindcasts and future projections of global inland and coastal nitrogen and phosphorus loads due to finfish aquaculture. *Reviews in Fisheries Science* 21, 112–156 (2013).

16. Durand, P. et al. Nitrogen processes in aquatic ecosystems. In *The European Nitrogen Assessment*, edited by M. A. Sutton et al., 7, 126–146 (Cambridge University Press, UK, 2011).

17. Peterson, B. J., Bahr, M. & Kling, G. W. A tracer investigation of nitrogen cycling in a pristine tundra river. *Canadian Journal of Fisheries and Aquatic Sciences* 54, 2361–2367 (1997).

18. Galloway, J. N. et al. The nitrogen cascade. *Bioscience* 53, 341–356 (2003).

19. Stocker, T. *Climate change 2013: the physical science basis: Working Group I contribution to the Fifth assessment report of the Intergovernmental Panel on Climate Change.* (Cambridge University Press, 2014).

20. O'Gorman, P. A. Sensitivity of tropical precipitation extremes to climate change. *Nature Geoscience* 5, 697–700 (2012).

21. Immerzeel, W. W., Van Beek, L. P. & Bierkens, M. F. Climate change will affect the Asian water towers. *Science* 328, 1382–1385 (2010).

22. Jeppesen, E. et al. Climate change effects on nitrogen loading from cultivated catchments in Europe: implications for nitrogen retention, ecological state of lakes and adaptation. *Hydrobiologia* 663, 1–21 (2011).

23. Baron, J. et al. The interactive effects of excess reactive nitrogen and climate change on aquatic ecosystems and water resources of the United States. *Biogeochemistry* 114, 71–92 (2013).

24. Kaushal, S. S. et al. Land use and climate variability amplify carbon, nutrient, and contaminant pulses: a review with management implications. *JAWRA Journal of the American Water Resources Association* 50, 585–614 (2014).

25. Mosley, L. M. et al. The impact of extreme low flows on the water quality of the Lower Murray River and Lakes (South Australia). *Water Resources Management* 26, 3923–3946 (2012).

26. Whitehead, P., Wilby, R., Battarbee, R., Kernan, M. & Wade, A. J. A review of the potential impacts of climate change on surface water quality. *Hydrological Sciences Journal* 54, 101–123 (2009).

27. Taylor, R. G. et al. Ground water and climate change. *Nature Climate Change* 3, 322–329 (2013).

28. Stuart, M., Gooddy, D., Bloomfield, J. & Williams, A. A review of the impact of climate change on future nitrate concentrations in groundwater of the UK. *Science of the Total Environment* 409, 2859–2873 (2011).

29. Suddick, E. & Davidson, E. The role of nitrogen in climate change and the impacts of nitrogen-climate interactions on terrestrial and aquatic ecosystems, agriculture, and human health in the United States: a technical report submitted to the US National Climate Assessment. *North American Nitrogen Center of the International Nitrogen Initiative (NANC-INI), Woods Hole Research Center* 149, 208 (2012).

30. Barnes, R. T., Williams, M. W., Parman, J. N., Hill, K. & Caine, N. Thawing glacial and permafrost features contribute to nitrogen export from Green Lakes Valley, Colorado Front Range, USA. *Biogeochemistry* **117**, 413–430 (2014).
31. O'Hare, M. T. et al. Eutrophication impacts on a river macrophyte. *Aquatic Botany* **92**, 173–178 (2010).
32. Moss, B., Jeppesen, E., Søndergaard, M., Lauridsen, T. L. & Liu, Z. Nitrogen, macrophytes, shallow lakes and nutrient limitation: resolution of a current controversy? *Hydrobiologia* **710**, 3–21 (2013).
33. Rosset, V. et al. Is eutrophication really a major impairment for small waterbody biodiversity? *Journal of Applied Ecology* **51**, 415–425 (2014).
34. Bornette, G. & Puijalon, S. Response of aquatic plants to abiotic factors: a review. *Aquatic Sciences* **73**, 1–14 (2011).
35. Shen, Q. et al. Beyond hypoxia: occurrence and characteristics of black blooms due to the decomposition of the submerged plant *Potamogeton crispus* in a shallow lake. *Journal of Environmental Sciences* **26**, 281–288 (2014).
36. Paerl, H. W., Hall, N. S. & Calandrino, E. S. Controlling harmful cyanobacterial blooms in a world experiencing anthropogenic and climatic-induced change. *Science of the Total Environment* **409**, 1739–1745 (2011).
37. Paerl, H. W. & Scott, J. T. Throwing fuel on the fire: synergistic effects of excessive nitrogen inputs and global warming on harmful algal blooms. *Environmental Science & Technology* **44**, 7756–7758 (2010).
38. Seitzinger, S. P. & Kroeze, C. Global distribution of nitrous oxide production and N inputs in freshwater and coastal marine ecosystems. *Global Biogeochemical Cycles* **12**, 93–113 (1998).
39. Reay, D. S., Smith, K. A. & Edwards, A. C. In *Biogeochemical investigations of terrestrial, freshwater, and wetland ecosystems across the globe*, 437–451 (Springer, 2004).
40. Burford, R. D. J. & Crees, R. Losses of nitrous oxide dissolved in drainage water from agricultural land. *Nature* **278**, 342–343 (1979).
41. Harrison, J. & Matson, P. Patterns and controls of nitrous oxide emissions from waters draining a subtropical agricultural valley. *Global Biogeochemical Cycles* **17** (2003).
42. Eggleston, S., Buendia, L., Miwa, K., Ngara, T. & Tanabe, K. IPCC guidelines for national greenhouse gas inventories. *Institute for Global Environmental Strategies, Hayama, Japan* (2006).
43. Nevison, C. Review of the IPCC methodology for estimating nitrous oxide emissions associated with agricultural leaching and runoff. *Chemosphere – Global Change Science* **2**, 493–500 (2000).
44. Williams, J. & Crutzen, P. Nitrous oxide from aquaculture. *Nature Geoscience* **3**, 143 (2010).
45. Garcia-Ruiz, R., Pattinson, S. & Whitton, B. Denitrification in river sediments: relationship between process rate and properties of water and sediment. *Freshwater Biology* **39**, 467–476 (1998).
46. Patrick, W. & Reddy, K. Fate of fertilizer nitrogen in a flooded rice soil. *Soil Science Society of America Journal* **40**, 678–681 (1976).
47. March, L. D. Permanent sedimentation of nitrogen, phosphorus, and organic carbon in a high arctic lake. *Journal of the Fisheries Board of Canada* **35**, 1089–1094 (1978).

48. Jansson, M., Andersson, R., Berggren, H. & Leonardson, L. Wetlands and lakes as nitrogen traps. *Ambio* **23**, 320–325 (1994).
49. Reddy, K., Patrick, W. & Broadbent, F. Nitrogen transformations and loss in flooded soils and sediments. *Critical Reviews in Environmental Science and Technology* **13**, 273–309 (1984).
50. Seitzinger, S. P. Denitrification in freshwater and coastal marine ecosystems: ecological and geochemical significance. *Limnology and Oceanography* **33**, 702–724 (1988).
51. Mitchell, M. J. et al. Climatic control of nitrate loss from forested watersheds in the northeast United States. *Environmental Science & Technology* **30**, 2609–2612 (1996).
52. Davidson, E. A. In *Biogeochemistry of global change*, 369–386 (Springer, 1993).
53. Bouwman, A., Boumans, L. & Batjes, N. Emissions of N2O and NO from fertilized fields: summary of available measurement data. *Global Biogeochemical Cycles* **16**, 6-1-6-13 (2002).
54. Reay, D. S., Edwards, A. C. & Smith, K. A. Importance of indirect nitrous oxide emissions at the field, farm and catchment scale. *Agriculture Ecosystems & Environment* **133**, 163–169, doi:10.1016/j.agee.2009.04.019 (2009).
55. Wilhelm, S. & Adrian, R. Impact of summer warming on the thermal characteristics of a polymictic lake and consequences for oxygen, nutrients and phytoplankton. *Freshwater Biology* **53**, 226–237 (2008).
56. Jankowski, T., Livingstone, D. M., Bührer, H., Forster, R. & Niederhauser, P. Consequences of the 2003 European heat wave for lake temperature profiles, thermal stability, and hypolimnetic oxygen depletion: implications for a warmer world. *Limnology and Oceanography* **51**, 815–819 (2006).
57. Elser, J. J., Marzolf, E. R. & Goldman, C. R. Phosphorus and nitrogen limitation of phytoplankton growth in the freshwaters of North America: a review and critique of experimental enrichments. *Canadian Journal of Fisheries and Aquatic Sciences* **47**, 1468–1477 (1990).
58. Dean, W. E. & Gorham, E. Magnitude and significance of carbon burial in lakes, reservoirs, and peatlands. *Geology* **26**, 535–538 (1998).
59. Einsele, G., Yan, J. & Hinderer, M. Atmospheric carbon burial in modern lake basins and its significance for the global carbon budget. *Global and Planetary Change* **30**, 167–195 (2001).
60. Bergström, A. K. & Jansson, M. Atmospheric nitrogen deposition has caused nitrogen enrichment and eutrophication of lakes in the northern hemisphere. *Global Change Biology* **12**, 635–643 (2006).
61. Heathcote, A. J. & Downing, J. A. Impacts of eutrophication on carbon burial in freshwater lakes in an intensively agricultural landscape. *Ecosystems* **15**, 60–70 (2012).
62. Reay, D. S., Dentener, F., Smith, P., Grace, J. & Feely, R. A. Global nitrogen deposition and carbon sinks. *Nature Geoscience* **1**, 430–437, doi:10.1038/ngeo230 (2008).
63. Schindler, D. W. The cumulative effects of climate warming and other human stresses on Canadian freshwaters in the new millennium. *Canadian Journal of Fisheries and Aquatic Sciences* **58**, 18–29, doi:10.1139/cjfas-58-1-18 (2001).
64. Huttunen, J. T. et al. Fluxes of methane, carbon dioxide and nitrous oxide in boreal lakes and potential anthropogenic effects on the aquatic

greenhouse gas emissions. *Chemosphere* **52**, 609–621, doi:10.1016/S0045-6535(03)00243-1 (2003).

65. Liikanen, A. & Martikainen, P. J. Effect of ammonium and oxygen on methane and nitrous oxide fluxes across sediment–water interface in a eutrophic lake. *Chemosphere* **52**, 1287–1293 (2003).

66. Glissman, K., Chin, K.-J., Casper, P. & Conrad, R. Methanogenic pathway and archaeal community structure in the sediment of eutrophic Lake Dagow: effect of temperature. *Microbial Ecology* **48**, 389–399 (2004).

67. Dacey, J., Drake, B. & Klug, M. Stimulation of methane emission by carbon dioxide enrichment of marsh vegetation. *Nature* **370**, 47–49 (1994).

68. Moss, B. et al. Allied attack: climate change and eutrophication. *Inland Waters* **1**, 101–105 (2011).

69. Mallin, M. A., Johnson, V. L., Ensign, S. H. & MacPherson, T. A. Factors contributing to hypoxia in rivers, lakes, and streams. *Limnology and Oceanography* **51**, 690–701 (2006).

70. Suddick, E. C., Whitney, P., Townsend, A. R. & Davidson, E. A. The role of nitrogen in climate change and the impacts of nitrogen–climate interactions in the United States: foreword to thematic issue. *Biogeochemistry* **114**, 1–10 (2013).

71. Johnson, P. T. et al. Linking environmental nutrient enrichment and disease emergence in humans and wildlife. *Ecological Applications* **20**, 16–29 (2010).

72. Spalding, R. F. & Exner, M. E. Occurrence of nitrate in groundwater – a review. *Journal of Environmental Quality* **22**, 392–402 (1993).

73. Pretty, J. N. et al. Environmental costs of freshwater eutrophication in England and Wales. *Environmental Science & Technology* **37**, 201–208 (2003).

74. Grizzetti, B. et al. Nitrogen as a threat to European water quality. In *The European Nitrogen Assessment,* edited by M. Sutton et al., 379–404 (Cambridge University Press, UK, 2011).

75. Powlson, D. S. et al. When does nitrate become a risk for humans? *Journal of Environmental Quality* **37**, 291–295 (2008).

76. McKnight, G., Duncan, C., Leifert, C. & Golden, M. Dietary nitrate in man: friend or foe? *British Journal of Nutrition* **81**, 349–358 (1999).

77. Ward, M. H. et al. Workgroup report: drinking-water nitrate and health-recent findings and research needs. *Environmental Health Perspectives* **113**, 1607–1614 (2005).

78. Hord, N. G., Tang, Y. & Bryan, N. S. Food sources of nitrates and nitrites: the physiologic context for potential health benefits. *The American Journal of Clinical Nutrition* **90**, 1–10 (2009).

79. Fewtrell, L. Drinking-water nitrate, methemoglobinemia, and global burden of disease: a discussion. *Environmental Health Perspectives* **112**, 1371–1374 (2004).

80. Weyer, P. J. et al. Municipal drinking water nitrate level and cancer risk in older women: the Iowa Women's Health Study. *Epidemiology* **12**, 327–338 (2001).

81. Butler, A. R. & Feelisch, M. Therapeutic uses of inorganic nitrite and nitrate from the past to the future. *Circulation* **117**, 2151–2159 (2008).

82. L'hirondel, J. *Nitrate and man: toxic, harmless or beneficial?* (CABI, 2002).

83. Weller, R. et al. Nitric oxide is generated on the skin surface by reduction of sweat nitrate. *Journal of Investigative Dermatology* **107**, 327–331 (1996).

84. Ignarro, L. J. After 130 years, the molecular mechanism of action of nitroglycerin is revealed. *Proceedings of the National Academy of Sciences of the United States of America* **99**, 7816–7817 (2002).

85. Lange, R. L. et al. Nonatheromatous ischemic heart disease following withdrawal from chronic industrial nitroglycerin exposure. *Circulation* **46**, 666–678 (1972).

86. SoRelle, R. Nobel prize awarded to scientists for nitric oxide discoveries. *Circulation* **98**, 2365–2366 (1998).

87. Fan, A. M. & Steinberg, V. E. Health implications of nitrate and nitrite in drinking water: an update on methemoglobinemia occurrence and reproductive and developmental toxicity. *Regulatory Toxicology and Pharmacology* **23**, 35–43 (1996).

88. Comly, H. H. Cyanosis in infants caused by nitrates in well water. *Journal of the American Medical Association* **129**, 112–116 (1945).

89. Avery, A. A. Infantile methemoglobinemia: reexamining the role of drinking water nitrates. *Environmental Health Perspectives* **107**, 583 (1999).

90. Sanchez-Echaniz, J., Benito-Fernández, J. & Mintegui-Raso, S. Methemoglobinemia and consumption of vegetables in infants. *Pediatrics* **107**, 1024–1028 (2001).

91. Savino, F. et al. Methemoglobinemia caused by the ingestion of courgette soup given in order to resolve constipation in two formula-fed infants. *Annals of Nutrition and Metabolism* **50**, 368–371 (2006).

92. Brunato, F., Garziera, M. G. & Briguglio, E. A severe methaemoglobinemia induced by nitrates: a case report. *European Journal of Emergency Medicine* **10**, 326–330 (2003).

93. Bruning-Fann, C. S. & Kaneene, J. The effects of nitrate, nitrite and N-nitroso compounds on human health: a review. *Veterinary and Human Toxicology* **35**, 521–538 (1993).

94. Ward, M. H. et al. Drinking water nitrate and the risk of non-Hodgkin's lymphoma. *Epidemiology* **7**, 465–471 (1996).

8
Marine Nitrogen and Climate Change

The three primary routes for entry of reactive nitrogen into the marine environment are in run-off from the land[1], as deposition from the atmosphere[2], and via nitrogen-fixing microbes, such as the cyanobacteria, in surface waters[3]. Run-off from land dominates inputs to most coastal waters, with atmospheric deposition and nitrogen fixation being more important in the open ocean. The world's rivers are estimated to supply over 20 million tonnes of inorganic nitrogen to estuaries and oceans each year, with three-quarters of this now being from human sources such as fertiliser and sewage[4]. A further 140 million tonnes of nitrogen enters the marine environment through nitrogen fixation, and around 50 million tonnes is deposited directly from the atmosphere in rain, snow and dust.

As on land and in freshwater systems, marine reactive nitrogen is cycled through various organic, inorganic and gaseous forms, with each atom of nitrogen potentially being incorporated into biomass, decomposed and mineralised many times before it is eventually converted to inert dinitrogen gas and lost to the atmosphere. The residence time of nitrogen in marine systems may be as little as minutes or hours – for instance, where nitrate flowing into an estuary is rapidly denitrified and emitted as nitrous oxide and dinitrogen gas. Conversely, some nitrogen may remain part of the marine nitrogen cycle for centuries or millennia, either being repeatedly recycled within the marine food chain or ending up bound in the particles of marine sediments for long periods[5].

In addition to the huge indirect disruption to the marine nitrogen cycle caused by agriculture and fossil fuel burning, the direct alteration of nitrogen flows through fishing and aquaculture impacts seas around the world. As large amounts of fish are taken from the sea for consumption on land, the nitrogen contained as proteins and amino acids in the tissues of the fish is also removed. For the 1960s it is estimated that the equivalent of 60 per cent of all the extra nitrogen added to the marine environment by human activities was returned to the land in the form of fish[6]. As marine nitrogen inputs have risen, this proportion has dropped to around 20 per cent, yet this transfer of marine nitrogen back to land in the form of fish still represents a major human-induced pathway in the global nitrogen cycle (Figure 8.1).

For aquaculture, intensive fish and shellfish farming in many coastal areas can result in large increases in reactive nitrogen inputs. The aquaculture sector is expanding rapidly, both in coastal seas and in freshwater systems around the world, with current reactive nitrogen inputs to the marine environment due to fish farming alone estimated to be over 300,000 tonnes per year[7]. Together with a boom in shellfish and seaweed aquaculture, these human-induced inputs to marine and freshwater systems are likely to keep on increasing[8].

One of the most overt natural transfers of marine nitrogen back to land is in the form of seabird guano, with the once-mountainous deposits of nitrogen-rich guano on the west coast of Peru testament to the cumulative importance of this removal pathway (Chapter 2). Such channelling of marine nitrogen back to land by seabirds, and to a lesser extent by mammals, can have significant local impacts where it is concentrated in one area. Large seabird colonies, for instance, can be major point sources of ammonia emissions, with the ammonia emitted from their guano fertilising areas of land or water downwind of the colony[9]. Likewise, the mass migration from sea to freshwater of some fish species, such as salmon, constitutes an important natural pathway for nitrogen export from marine systems[10,11].

Climate forcing and carbon fluxes

The most direct impact of marine nitrogen on global climate change is in the emission of nitrous oxide into the atmosphere. Most nitrous oxide is formed via either denitrification or nitrification by microbes[12], with the former being predominant in low-oxygen waters and sediments[13] and the latter being more important in well-oxygenated surface waters[14]. Because of the high reactive nitrogen loadings, shallow depths and fast processing rates common to estuaries and continental shelf waters[15] (those less than ~200 m deep), these areas tend to dominate nitrous oxide emissions from the marine environment[16]. In deeper ocean waters large amounts of nitrous oxide may also be produced, but a

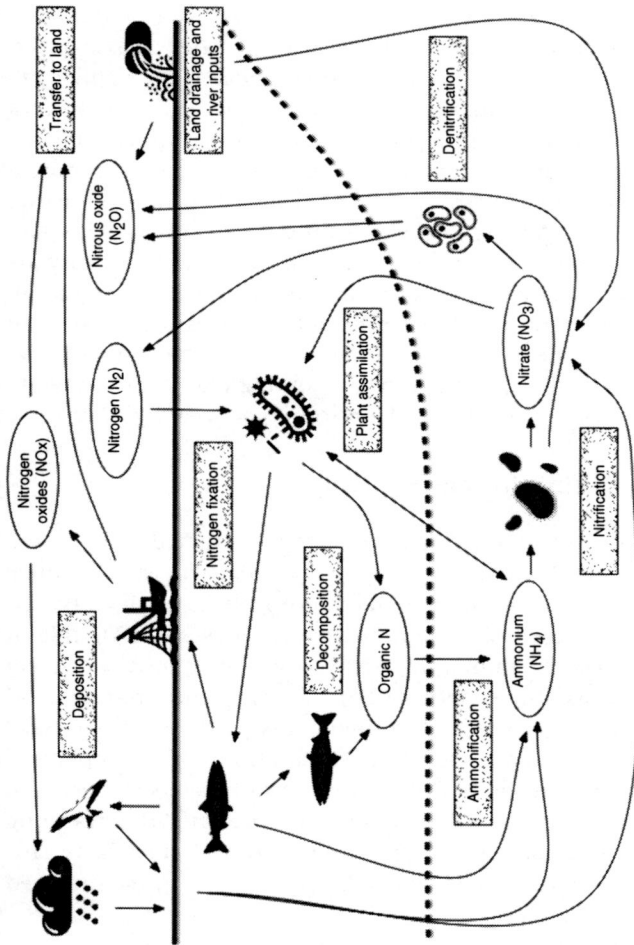

Figure 8.1 The marine nitrogen cycle

The solid line represents the ocean surface and the dotted line the ocean floor. The boxes show the key processes by which nitrogen is cycled and the ovals show the changing form of nitrogen as it undergoes these processes. Denitrification is a major process for converting the reactive nitrogen to inert dinitrogen gas (N_2) in many marine systems. Anaerobic ammonium oxidation (called 'anammox') is also very important in marine environments, with the anammox bacteria able to convert the ammonium and nitrite directly to dinitrogen.

Source: Dave Reay

significant proportion of this is usually reduced to dinitrogen gas before it can escape into the atmosphere[17].

Globally, estuaries and continental shelf seas are estimated to emit around one million tonnes (Tg) of nitrogen in the form of nitrous oxide each year[4]. In terms of its climate-warming effect, one million tonnes of nitrogen as nitrous oxide is equivalent to releasing over 450 million tonnes of carbon dioxide. Though these areas are often the strongest individual sources, their relatively small expanse means they only comprise about 20 per cent of marine nitrous oxide emissions, with most of the rest – about four million tonnes of nitrogen as nitrous oxide per year – coming from the open ocean. For the total marine environment then, current nitrous oxide emissions are estimated to be around five million tonnes of nitrogen per year[2,18]. The intensifying inputs of human-induced nitrogen deposition to the oceans during the 20th century are thought to have already boosted marine nitrous oxide emissions by about 1.6 million tonnes a year. In climate-forcing terms, this is equivalent to the total carbon dioxide emissions from France and Italy combined. Yet more anthropogenic nitrogen inputs in the coming years are expected to enhance emissions by almost two million tonnes per year by 2030[2].

Nitrogen and the marine carbon sink

The long-term ocean sink for carbon dioxide is estimated to be in the range of about 1.5 to 2 billion tonnes (Pg) of carbon per year. This huge amount is equivalent to around one-quarter of global carbon emissions from fossil fuel burning[19–21]. As such, the oceans play a leading role in buffering the impact of human-induced carbon emissions on global climate. However, as atmospheric carbon dioxide concentrations and global temperatures continue to rise, the ability of the oceans to take up and store ever-increasing amounts of carbon from the atmosphere is likely to weaken[22,23]. Though the ocean carbon sink is expected to keep on expanding for much of the 21st century, it is predicted that warming waters, changes in circulation and increased stratification of the oceans will mean that this expansion will fail to keep pace with human-induced carbon dioxide emissions.

As a larger proportion of human-induced carbon dioxide emissions will therefore remain in the atmosphere, overall global warming will increase even further – a so-called 'positive feedback'[24,25].

Just as with terrestrial ecosystems, changing nitrogen inputs to the oceans have the potential to radically alter carbon cycling in marine ecosystems, with potentially major consequences both for global carbon dioxide concentrations and for marine biodiversity[26]. The amount of reactive nitrogen entering marine systems is expected to increase by around 50 per cent by 2100[27] as nitrogen emissions from fossil fuel burning and agriculture expand in many areas of the world. These growing inputs may boost plant growth and ocean carbon uptake in the coming decades and, if the net increase in primary production is large enough, may help to bolster the ocean carbon sink in the face of 21st-century warming.

Unfortunately, the response of the ocean carbon sink to rising nitrogen inputs appears rather limited, and more reactive nitrogen in our oceans will also bring with it a climate penalty in the form of enhanced nitrous oxide emissions[28]. In a future with relatively high nitrogen emissions and deposition to the oceans, primary production is expected to be boosted by up to one billion tonnes of carbon per year by 2100, but only about 160 million tonnes of this would be locked away in deep ocean waters and sediments for a significant period. This slight enhancement of the marine carbon sink would provide a reduction in atmospheric carbon dioxide concentrations of just 1.66 parts per million by the end of the century – less than the growth in carbon dioxide concentrations we are currently seeing every year[29]. With the search for solutions to human-induced climate change becoming increasingly desperate, a host of 'silver bullet' strategies – including deliberate nitrogen additions – to engineer much larger increases in ocean algal growth and carbon uptake have been suggested (see Chapter 12).

Climate change and marine nitrogen processing

As water temperatures increase due to the enhanced greenhouse effect, the rate of nitrogen processing in many marine systems is also likely to increase, with a faster turnover of its organic, mineralised and gaseous forms[30]. As higher temperatures will increase decomposition and respiration rates[31], a corresponding expansion of low-oxygen conditions in estuarine and coastal seawater is expected[32]. Such conditions may lead to an increase in denitrification and thereby result in increasing nitrous oxide emissions into the atmosphere[33,34].

Higher temperatures and changing wind speeds may also increase stratification of marine waters in some regions. In the open ocean in particular, more stratification of surface waters may limit the supply of reactive nitrogen for algal growth and so reduce primary production and carbon uptake[25]. These changes could also lead to major changes in algal community structure in favour of nitrogen-fixing species. For shallower waters and enclosed seas such as the Gulf of Mexico, increased stratification risks enhancing low-oxygen and anoxic conditions in bottom waters[35] (Figure 8.2).

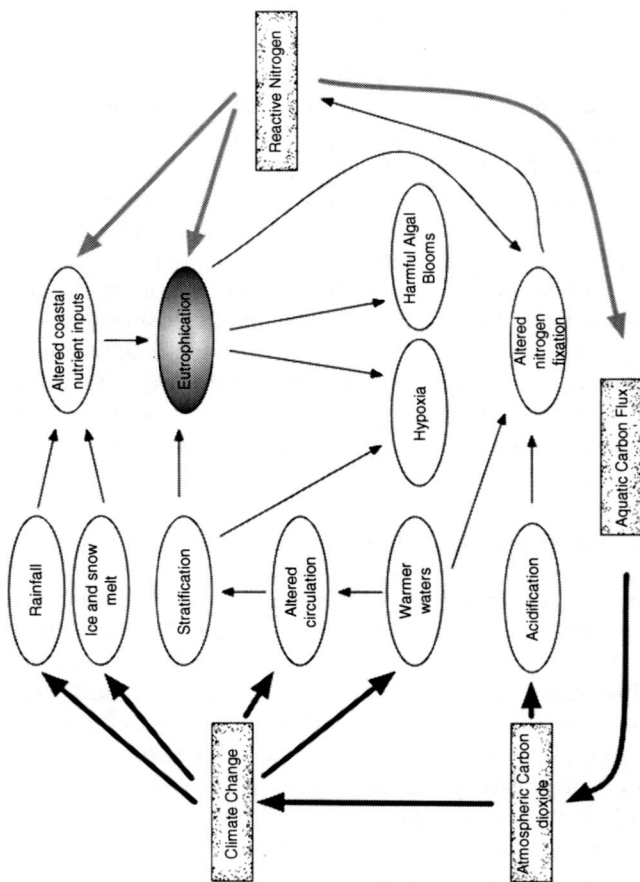

Figure 8.2　Key interactions of nitrogen and climate change in marine environments

The major impacts of climate change and increasing carbon dioxide concentrations (large black arrows) and the major impacts of nitrogen (large grey arrows) come together directly in the marine environment through effects such as eutrophication (from more nitrogen inputs and increased run-off). Harmful algal blooms and hypoxia (low oxygen) impacts are especially important in coastal environments such as the Gulf of Mexico.

Source: Dave Reay

The increases in intense rainfall events projected in many areas for the 21st century will result in higher nitrogen loads being flushed into estuarine and coastal systems, creating periods of very high freshwater, sediment and nitrogen inputs, as well as boosting overall nitrogen loads each year[36,37]. More reactive nitrogen supply in such large pulses may increase the amount transported beyond the confines of estuaries and out into coastal and offshore waters[38]. As a result, eutrophication and harmful algal blooms in these areas may become more common[39]. For areas that experience sharp reductions in precipitation, lower water flows into estuaries and coastal waters will allow more of the nitrogen to be processed there before it escapes to the open ocean[40]. This could increase marine plant and algal growth and also enhance estuarine nitrous oxide emissions via denitrification[41]. Likewise, the combination of more decomposing organic material and widespread low-oxygen conditions that result from eutrophication has the potential to enhance methane emissions from many marine environments, estuarine and coastal systems in particular[16].

Finally, the acidification of the oceans resulting from increasing carbon dioxide concentrations in the atmosphere may further alter algal communities[42], nitrogen availability and processes. A lower pH would restrict nitrification[43], and so could reduce nitrous oxide emissions from this source. However, this effect may be offset by the expansion of low-oxygen conditions – and denitrification – in many ocean areas. The true impact of ocean acidification on the marine nitrogen cycle and its climate-change interactions remains highly uncertain[44].

Harmful algal blooms

Algal growth in the surface waters of many oceanic areas is held back by the supply of nutrients for growth[45,46], with nitrogen or iron limitation being common in the open ocean[47] and a combination of nitrogen and phosphorus limitation characteristic of coastal and estuarine areas[48]. As increasingly fertile winds and rivers dump their plant-boosting loads into ocean waters around the world, the algal blooms that result are growing larger and more widespread. One of the biggest of these blooms, which returns each year with devastating results, is in the Gulf of Mexico and is known as a 'dead zone'[49].

Dead zones existed long before Haber and Bosch first pulled on their lab coats. These zones can start to develop wherever poorly mixed coastal waters coincide with lots of decomposing plant material. Humankind is helping to boost their number, size and intensity by fertilising the oceans with vast amounts of extra nitrogen and phosphorus[50]. In the warm, sun-drenched waters of the Gulf of Mexico algae can respond quickly to the rich nitrogen inputs that flow down from the intensive farmlands of America's Midwest[51,52]. Many microscopic algae are able to double their numbers every 12 hours, so that within one week a single cell floating in clear nutrient-enriched waters can multiply into more than 16,000 new ones. As winter storms give way to the calm of spring and summer in the Gulf of Mexico, the bloom begins to spread through the stagnating waters. At its peak it stretches across thousands of square miles and is clearly visible from space as a green pall billowing along the coastline. The exact timing, severity and composition of the blooms are hard to predict, but when they strike the effects are all too obvious.

For the trillions of cells within the bloom, life is both luxurious and short. When they die and sink towards the seabed, myriads of bacteria get to work on breaking them down. As the microbes feast on the disintegrating clouds of algae, they draw more and more oxygen from the water around them to power the frenzy of decomposition. The larger and more intense the bloom, the bigger the dieback. In the poorly mixed depths of the Gulf of Mexico the water becomes so starved of oxygen that any animals too slow to move clear quickly join the spreading soup of decomposing dead. Shellfish and sea anemones, starfish and urchins, all are obliterated as the suffocating water overtakes them. In addition to that produced by the huge Gulf of Mexico bloom, dead zones are now common in the Black Sea, the Baltic and in the coastal waters of China, Japan and southeast Australia[53,54].

Even those animals that are able to retreat from the dead zone can be drastically affected, and their food supplies and habitats destroyed by the flush of algae that deprives the ocean floor of both oxygen and light. Fish and shrimp stocks may crash, as the protective nurseries previously provided by swaying stands of sea grasses are reduced to shadowy stretches of barren mud[55]. In death then, algal blooms can threaten marine life around the world, but a small percentage of the many thousands of algal species in the oceans pose a much more direct threat to the lives of fish, birds, sea mammals and even humans (Box 8.1).

Harmful algal blooms and human health

Of the thousands of known algal species, fewer than a hundred types are classed as toxic[57]. Where such poisonous forms do occur, they are usually present in numbers too low to pose much of a direct hazard to humans. It is in bloom conditions that the amounts of toxin they

Box 8.1

Fish kills and *Pfiesteria*

Each year, along the coast of North Carolina millions of Atlantic Menhanden – a silver-coloured, herring-like fish – come inshore to feast on the abundant algae. As the shoals gather, the waters become thick with their secretions, and in the muddy sea floor a deadly predator awakes. Emerging from their well-protected cyst stage, billions of microorganisms called *Pfiesteria*[56] swarm up to ambush the unsuspecting shoals. For much of the time these remarkable microbes feast only on bacteria and algae in the water, quickly responding to any flush of new nitrogen and the algal smorgasbord that develops. If times are hard, they can even enslave the chloroplasts contained in the algae they have eaten to provide a few days of photosynthesis, before adopting the cyst form and drifting back down to the seabed to wait for the Menhanden to arrive.

Quick to react to the chemical signals that fish are present, the *Pfiesteria* attach themselves to fish skin, gills and mouths. There they begin to strip away the surface layers of tissue, their vast numbers overwhelming the fish and opening expanding legions in their flesh. Within 24 hours of commencing the attack most victims are dead, their killers continuing to feast on the decomposing remains until food supplies are exhausted and it is time to retreat to the depths once again. As waters become warmer and nitrogen inputs increase, the range of these hit-and-run killers is likely to expand and the number of such devastating ambush attacks to increase.

Some large fish kills associated with algal blooms are less deliberate, the algae being of a type with long spines that clog gills and cause suffocation. Tracking down the culprit after the dead fish begin to wash ashore can take an awful lot of detective work, the vanishing act of killers like *Pfiesteria* making things especially difficult. As with dead zones, fish kills linked to algal blooms can be traced back to well before human activities began to add large amounts of nitrogen to the oceans. It is in the growing number and range of such incidents that extra nitrogen has been implicated. Stinking tides of dead fish, empty nets and bankrupt fishermen are an obvious penalty for this profligacy, but harmful blooms demand the ultimate price from many hundreds of people each year.

produce can reach dangerous levels. Reports of toxic algal blooms go back hundreds of years, yet in recent decades their number and toll on human health seems to have risen sharply (Box 8.2). More than 60,000 people now go down with some form of algal toxin poisoning each year. For most this constitutes little more than a bad stomach upset and a few uncomfortable days within dashing distance of a toilet; for around 900 of those poisoned the consequences are fatal[58].

The beaches of the world sport warning signs for everything from rip tides and quicksand to jellyfish and sharks, keeping safe the many millions who flock to the seaside for summer sun and sandcastles, which is vital for countless coastal town economies. Precautions like lifeguards and shark nets might help keep these money-spinning tourist magnets open, but when algal blooms creep towards the shore the damage to both businesses and people can be crippling.

Box 8.2

Toxic algae

On the face of it, the small coastal town of Capitola in central California is little different from hundreds of such towns scattered along the western seaboard of the US. It has a small harbour and a nice beach and plays host to an annual begonia festival. One warm summer's day in 1961, Capitola became the centre of a mystery that would take almost 30 years to solve. Out of the sky that morning came a huge flock of seabirds. Hundreds of sooty shearwaters – shy fish-eating birds that normally come ashore only to breed – dive-bombed the town. They dashed themselves into cars and lamp posts, smashed through windows, vomited half-digested anchovies on the sidewalks and attacked the fleeing townspeople.

Unsurprisingly, the avian mayhem of Capitola helped to inspire Alfred Hitchcock's horror film 'The Birds'. At the time it was thought that sea fog had made the birds disorientated, but the fish vomiting suggested a very different culprit. Decades later the prime suspects for this bird attack and for a growing number of unexplained sea-food poisoning incidents around the world began to emerge. These agents of delirium and death were tracked down via the toxic calling cards they left behind[59,60]. They were microscopic, occurred in myriad forms and were causing increasing problems in lakes, rivers and oceans around the world. They were algae.

At best, a large bloom coming ashore will mean unsightly bathing waters and malodorous surface scums that send the tourists streaming back inland with car windows tightly shut. At worst, the bloom will be exuding powerful toxins that make even a seafront stroll perilous[61,62].

Early in 2005 the first signs of a red tide – an intense bloom of toxic red algae – were seen in the waters off the coast of Florida. Fish began to die, washing ashore in increasing numbers. Turtles and dolphins also began to succumb, and soon the discoloured plumes of ocean water had closed in on the beaches. Hospital admissions soared as toxin-laden sea breezes swept onshore to inflame the eyes, throats and lungs of those exposed. Recovery from such sea spray poisoning is usually swift, the doses tending to be low[63]. Those who have direct contact with contaminated water, such as swimmers, run the risk of more severe effects, and there have been several recorded incidents of pet dogs receiving lethal doses by licking water off their coats after a swim.

By midsummer that year some 15,000 square miles of the US coastline had been closed and there were losses of more than $10 million in shellfish sales in Maine and Massachusetts alone. Closing beaches and staying indoors can help reduce the direct human health costs, albeit with severe consequences for the local economy. Yet there is often a longer-lived sting in the tail of such blooms[64].

Shellfish poisoning

Shellfish get their food by filtering the water around them and trapping any algae floating in it. If these algae are toxic, the poisons held within their tissues may build up to lethal levels over time. When these shellfish are eaten they deliver a concentrated dose straight to the stomach of the eater. Cooking shellfish is widely held as a good way of avoiding food poisoning, yet many of the toxins that algae produce are resistant to heat. As scores of diners in the restaurants of Canada's Prince Edward Island found to their cost one dark week in the winter of 1987, an otherwise healthy plate full of cooked shellfish can hide a world of pain[65].

The concentrated dose of toxins served up to the unsuspecting inhabitants of Prince Edward Island was in a batch of farmed blue mussels. Within hours of their meal more than a hundred people had become gravely ill. Some began to vomit and complain of cramps and headaches, others became aggressive or broke down in tears. As the toxins spread through their bodies, increasing numbers became disorientated and suffered amnesia. The most severely affected were wracked by seizures and lapsed into comas. Three of the victims died and of the rest around one-quarter suffered memory loss. Some were unable to recognise close relatives or their own homes, while many had difficulty in recalling anything that had happened since the meal. For a dozen of the survivors the memory loss proved permanent[66].

The tragic results of this mass poisoning led to it being given the tag of 'amnesic shellfish poisoning', but exactly what had caused the brain damage remained unclear. Never before had food poisoning been associated with such severe neurological effects, and it was only through a drawn-out process of systematic testing that the team of marine biologists and chemists hastily assembled to work on the problem were able to identify the true culprit[67]. Viruses and bacteria were the first to be exonerated, the tests for the usual suspects like heavy metals and pesticides drew a blank, and finally came a series of experiments involving the injection of different samples from the contaminated blue mussels into lab mice. If the mice failed to react, that fraction could be ruled out, but if the mice started scratching their shoulders with their hind legs – a sign of damage to their nervous systems – then the team could zero in on the guilty toxin. They finally identified it as domoic acid, a poison that induces a cascade of over-excitement and death in nerve cells. The algae responsible had left their calling card in pigments extracted from the tissues of the blue mussels, and the same algae-borne domoic acid was quickly implicated as the poison in the anchovies that had caused the bird-vomiting mayhem of Capitola several decades before[68].

This deadly toxin has now been detected in many coastal areas of the world and implicated in the deaths of hundreds of seabirds and sea lions around the US. Thanks to stringent screening of seafood following the Prince Edward Island outbreak, only a few new cases of amnesic shellfish poisoning in humans have been reported[69]. Yet there are a host of other algal toxins that can slip through the screening net.

These other toxic groupings include diarrhetic shellfish poisoning, neurotoxic shellfish poisoning and paralytic shellfish poisoning. As their names suggest, none of them are pleasant. The diarrhetic type is most common in the waters around Europe and Japan and has the familiar food poisoning symptoms of vomiting, muscle aches and diarrhea[70]. For neurotoxic poisoning – associated with the eye-stinging toxins swept ashore from red tides – the effects ratchet up somewhat, often adding dizziness, numbness and a reversal of the sensations of heat and cold to the diarrhetic mix[71]. Though highly unpleasant, most victims fully recover within a few days, and no deaths have been recorded. Paralytic shellfish poisoning can be in a different league entirely.

Cases of paralytic shellfish poisoning can be traced back at least 200 years and, along with its diarrhetic and neurotoxic cousins, the number and global spread of these poisoning incidents appears to be rising rapidly. The neurotoxin responsible is one of the most powerful in existence, and along with the paralysis that gives it its name, it can cause slurred speech, drowsiness and burning sensations[72]. For the crew members taking their seats around the dinner table of a small fishing boat off the coast of Nantucket in the summer of 1990, it was blue mussels that again served as the vehicle for concentrated algal toxins[73]. The mussels had been dragged up with the nets earlier in the day, and the cook had then boiled them for over an hour, serving them up with rice, potatoes and a green salad. The boat's captain was late to the table, his men having already stuck into their eagerly awaited dinner. It wasn't long before it became clear that something was very wrong. One by one the crew began to complain of numb mouths and faces, they started to vomit and four of them lost all feeling in their limbs. With his men fast deteriorating and paralysis beginning to creep through his own body, the captain sent out a distress call to the coast guard. They were lucky. Rushed by helicopter to the local hospital's emergency room, all of them survived to fish again.

There is no antidote to the toxins that cause paralytic shellfish poisoning, and in severe cases these toxins will quickly lead to suffocation of the victim. This toxin kills roughly one in six of those affected, and just three years before the lucky escape of the six fishermen – when mussel-borne domoic acid was wiping the memories of the diners in Prince Edward Island – these paralysing toxins were deadening the tissues of scores of people in Guatemala. On that occasion more than 180 people had eaten contaminated clams. For 26 of them the meal was fatal[74].

Early warning and catchment management

Almost every coastal country in the world is now threatened by harmful algal blooms, and in the US alone costs amount to millions of dollars each year[75]. Recent decades have seen a steep rise in the number being reported and the damage caused, although the direct link to more nitrogen being pumped into the oceans is still controversial[76–79]. Though nitrogen is a key nutrient for algal growth, other factors – such as the temperature of the water, how well mixed it is and, most importantly, where exactly the bloom ends up – can all help determine just how big a problem develops. Records of blooms and outbreaks of toxic shellfish poisoning dating back to the time of Captain Cook prove that they can occur without the helping hand of the fertiliser industry[80]. Indeed, many of the most harmful blooms are now known to form offshore, well away from the enriching rivers of nitrogen that drain the fields and towns inland. It is when wind and tide combine to sweep them back towards land that their impacts on humans really start to hit home, the extra boost supplied to them by the flush of new nutrients allowing them to expand to harmful proportions.

The waters along the coast of Florida have been receiving ever-greater amounts of nitrogen since the 1950s, as the dense nutrient-filtering Everglades have been bypassed or converted for farming and buildings[81]. At the same time the number of toxic red tide algae being found along these coasts has seen a 15-fold increase. It seems then that while harmful algal blooms may well develop entirely naturally, heaping more and more nitrogen into oceans around the world can make them a good deal worse[82]. Numerous studies have now reported such a link, and efforts to better predict where and when blooms will occur are now being focussed on the nutrients supplied to the oceans from both land and air. These predictions are still far from perfect, relying on a complex combination of weather forecasts, fishery reports, water sampling and satellite detection of the small changes in ocean colour that hint that a big bloom might be on its way[83–85].

In addition to better prediction, efforts to try and cut the number and severity of these blooms are increasingly targeting the huge nitrogen and phosphorous losses coming from the land. Cutting off these enriching supplies may help to prevent some of the blooms entirely and at least reduce the size of others[86,87]. In recent years a big push to tackle the huge dead zone in the Gulf of Mexico has involved the construction of wetlands to intercept the nutrient-rich water seeping from the farmlands and sewage works of the Southern US before it can get to the sea[88]. A key target has been nitrate – the highly mobile form of nitrogen that slips so easily from farmland soils and moves downstream towards the hungry clouds of algae. Yet even large-scale efforts such as this may achieve little in the face of a changing global climate. The altered weather patterns and warming of our seas that are predicted for coming decades could help bigger and more damaging blooms to develop[89]. This, along with a growing human population concentrated around the world's coastlines, is likely to mean that the harmful algal bloom problem will get worse before it gets better.

References

1. Kroeze, C., Seitzinger, S. P. & Domingues, R. Future trends in worldwide river nitrogen transport and related nitrous oxide emissions: a scenario analysis. *The Scientific World Journal* **1 Suppl 2**, 328–335, doi:10.1100/tsw.2001.279 (2001).
2. Duce, R. A. et al. Impacts of atmospheric anthropogenic nitrogen on the open ocean. *Science* **320**, 893–897, doi:10.1126/science.1150369 (2008).

3. Deutsch, C., Sarmiento, J. L., Sigman, D. M., Gruber, N. & Dunne, J. P. Spatial coupling of nitrogen inputs and losses in the ocean. *Nature* **445**, 163–167 (2007).
4. Seitzinger, S. P. & Kroeze, C. Global distribution of nitrous oxide production and N inputs in freshwater and coastal marine ecosystems. *Global Biogeochemical Cycles* **12**, 93–113 (1998).
5. Zehr, J. P. & Ward, B. B. Nitrogen cycling in the ocean: new perspectives on processes and paradigms. *Applied and Environmental Microbiology* **68**, 1015–1024 (2002).
6. Maranger, R., Caraco, N., Duhamel, J. & Amyot, M. Nitrogen transfer from sea to land via commercial fisheries. *Nature Geoscience* **1**, 111–112 (2008).
7. Bouwman, A. F. et al. Hindcasts and future projections of global inland and coastal nitrogen and phosphorus loads due to finfish aquaculture. *Reviews in Fisheries Science* **21**, 112–156 (2013).
8. Voss, M. et al. Nitrogen processes in coastal and marine ecosystems. *The European Nitrogen Assessment: Sources, Effects and Policy Perspectives* **1**, 147–176 (2011).
9. Riddick, S. et al. The global distribution of ammonia emissions from seabird colonies. *Atmospheric Environment* **55**, 319–327 (2012).
10. Rex, J. F. & Petticrew, E. L. Delivery of marine-derived nutrients to streambeds by Pacific salmon. *Nature Geoscience* **1**, 840–843 (2008).
11. Schindler, D. E. et al. Pacific salmon and the ecology of coastal ecosystems. *Frontiers in Ecology and the Environment* **1**, 31–37 (2003).
12. Bange, H. W., Freing, A., Kock, A. & Löscher, C. Marine pathways to nitrous oxide. In *Nitrous Oxide and Climate Change*, edited by K. Smith, 36–54 (Earthscan, New York, 2010).
13. Gruber, N. & Sarmiento, J. L. Global patterns of marine nitrogen fixation and denitrification. *Global Biogeochemical Cycles* **11**, 235–266 (1997).
14. Freing, A., Wallace, D. W. & Bange, H. W. Global oceanic production of nitrous oxide. *Philosophical Transactions of the Royal Society B: Biological Sciences* **367**, 1245–1255 (2012).
15. Herbert, R. Nitrogen cycling in coastal marine ecosystems. *FEMS Microbiology Reviews* **23**, 563–590 (1999).
16. Bange, H. W. Nitrous oxide and methane in European coastal waters. *Estuarine, Coastal and Shelf Science* **70**, 361–374 (2006).
17. Cohen, Y. & Gordon, L. I. Nitrous oxide in the oxygen minimum of the eastern tropical North Pacific: evidence for its consumption during denitrification and possible mechanisms for its production. *Deep Sea Research* **25**, 509–524 (1978).
18. Smith, K., Crutzen, P., Mosier, A. & Winiwarter, W. The global nitrous oxide budget: a reassessment. In *Nitrous Oxide and Climate Change*, edited by K. Smith, 63–84 (Earthscan, New York, 2010).
19. Sabine, C. L. et al. The oceanic sink for anthropogenic CO2. *Science* **305**, 367–371 (2004).
20. Le Quéré, C., Raupach, M. R., Canadell, J. G. & Marland, G. Trends in the sources and sinks of carbon dioxide. *Nature Geoscience* **2**, 831–836 (2009).
21. Cox, P. M., Betts, R. A., Jones, C. D., Spall, S. A. & Totterdell, I. J. Acceleration of global warming due to carbon-cycle feedbacks in a coupled climate model. *Nature* **408**, 184–187 (2000).

22. Sarmiento, J. L. & Le Quéré, C. Oceanic carbon dioxide uptake in a model of century-scale global warming. *Science* **274**, 1346–1350 (1996).
23. Le Quéré, C. et al. Saturation of the southern ocean CO2 sink due to recent climate change. *Science (New York, NY)* **316**, 1735–1738 (2007).
24. Canadell, J. G. et al. Contributions to accelerating atmospheric CO2 growth from economic activity, carbon intensity, and efficiency of natural sinks. *Proceedings of the National Academy of Sciences of the United States of America* **104**, 18866–18870 (2007).
25. Fung, I. Y., Doney, S. C., Lindsay, K. & John, J. Evolution of carbon sinks in a changing climate. *Proceedings of the National Academy of Sciences of the United States of America* **102**, 11201–11206 (2005).
26. Krishnamurthy, A., Moore, J. K., Zender, C. S. & Luo, C. Effects of atmospheric inorganic nitrogen deposition on ocean biogeochemistry. *Journal of Geophysical Research: Biogeosciences (2005–2012)* **112** (2007).
27. Galloway, J. N. The global nitrogen cycle: past, present and future. *Science in China. Series C, Life sciences/Chinese Academy of Sciences* **48 Suppl 2**, 669–678, doi:10.1007/BF03187108 (2005).
28. Reay, D. S., Dentener, F., Smith, P., Grace, J. & Feely, R. A. Global nitrogen deposition and carbon sinks. *Nature Geoscience* **1**, 430–437, doi:10.1038/ngeo230 (2008).
29. Keeling, C., Whorf, T., Wahlen, M. & Plicht, J. v. d. Interannual extremes in the rate of rise of atmospheric carbon dioxide since 1980. *Nature* **375**, 666–670 (1995).
30. Riebesell, U., Körtzinger, A. & Oschlies, A. Sensitivities of marine carbon fluxes to ocean change. *Proceedings of the National Academy of Sciences of the United States of America* **106**, 20602–20609 (2009).
31. Doney, S. C. et al. Climate change impacts on marine ecosystems. *Marine Science* **4**, 11–37 (2012).
32. Gruber, N. Warming up, turning sour, losing breath: ocean biogeochemistry under global change. *Philosophical Transactions of the Royal Society A: Mathematical, Physical and Engineering Sciences* **369**, 1980–1996 (2011).
33. Codispoti, L. et al. The oceanic fixed nitrogen and nitrous oxide budgets: moving targets as we enter the anthropocene? *Scientia Marina* **65**, 85–105 (2001).
34. Naqvi, S. et al. Increased marine production of N2O due to intensifying anoxia on the Indian continental shelf. *Nature* **408**, 346–349 (2000).
35. Keeling, R. F., Körtzinger, A. & Gruber, N. Ocean deoxygenation in a warming world. *Marine Science* **2** (2010).
36. Rabalais, N. N., Turner, R. E., Díaz, R. J. & Justić, D. Global change and eutrophication of coastal waters. *ICES Journal of Marine Science: Journal du Conseil* **66**, 1528–1537 (2009).
37. Voss, M. et al. The marine nitrogen cycle: recent discoveries, uncertainties and the potential relevance of climate change. *Philosophical Transactions of the Royal Society B: Biological Sciences* **368**, 20130121 (2013).
38. Howarth, R. et al. Coupled biogeochemical cycles: eutrophication and hypoxia in temperate estuaries and coastal marine ecosystems. *Frontiers in Ecology and the Environment* **9**, 18–26 (2011).
39. Paerl, H. W. & Huisman, J. Climate change: a catalyst for global expansion of harmful cyanobacterial blooms. *Environmental Microbiology Reports* **1**, 27–37 (2009).

40. Howarth, R. W., Swaney, D. P., Butler, T. J. & Marino, R. Rapid communi-cation: climatic control on eutrophication of the Hudson River Estuary. *Ecosystems* 3, 210–215 (2000).
41. Kroeze, C. & Seitzinger, S. P. Nitrogen inputs to rivers, estuaries and con-tinental shelves and related nitrous oxide emissions in 1990 and 2050: a global model. *Nutrient Cycling in Agroecosystems* 52, 195–212 (1998).
42. Boyd, P. W. & Doney, S. C. Modelling regional responses by marine pelagic ecosystems to global climate change. *Geophysical Research Letters* 29, 53-51-53-54 (2002).
43. Beman, J. M. et al. Global declines in oceanic nitrification rates as a conse-quence of ocean acidification. *Proceedings of the National Academy of Sciences of the United States of America* 108, 208–213 (2011).
44. Hutchins, D. A., Mulholland, M. R. & Fu, F. Nutrient cycles and marine microbes in a CO2-enriched ocean. *Oceanography* 22, 128–145 (2009).
45. Howarth, R. W. Nutrient limitation of net primary production in marine ecosystems. *Annual Review of Ecology and Systematics* 19, 89–110 (1988).
46. Vitousek, P. M. & Howarth, R. W. Nitrogen limitation on land and in the sea: how can it occur? *Biogeochemistry* 13, 87–115 (1991).
47. Behrenfeld, M. J., Bale, A. J., Kolber, Z. S., Aiken, J. & Falkowski, P. G. Confirmation of iron limitation of phytoplankton photosynthesis in the equatorial Pacific Ocean. *Nature* 383, 508–511 (1996).
48. Conley, D. J. et al. Controlling eutrophication: nitrogen and phosphorus. *Science* 323, 1014–1015 (2009).
49. Rabalais, N. N., Turner, R. E. & Wiseman Jr, W. J. Gulf of Mexico hypoxia, AKA "The Dead Zone". *Annual Review of Ecology and Systematics* 33, 235–263 (2002).
50. Dodds, W. K. Nutrients and the "Dead Zone": the link between nutrient ratios and dissolved oxygen in the northern Gulf of Mexico. *Frontiers in Ecology and the Environment* 4, 211–217 (2006).
51. Rabalais, N. N., Turner, R. E. & Wiseman, W. J. Hypoxia in the Gulf of Mexico. *Journal of Environmental Quality* 30, 320–329 (2001).
52. Burkart, M. R. & James, D. E. Agricultural-nitrogen contributions to hypoxia in the Gulf of Mexico. *Journal of Environmental Quality* 28, 850–859 (1999).
53. Diaz, R. J. & Rosenberg, R. Spreading dead zones and consequences for marine ecosystems. *Science* 321, 926–929 (2008).
54. Diaz, R. J. Overview of hypoxia around the world. *Journal of Environmental Quality* 30, 275–281 (2001).
55. Craig, J. K. et al. Ecological effects of hypoxia on fish, sea turtles, and marine mammals in the northwestern Gulf of Mexico. In *Coastal Hypoxia: Consequences for Living Resources and Ecosystems*, edited by N. Rabalais et al., 58, 269–291 (AGU Coastal and Estuarine Studies Series, USA, 2001).
56. Burkholder, J. M. & Marshall, H. G. Toxigenic *Pfiesteria* species – updates on biology, ecology, toxins, and impacts. *Harmful Algae* 14, 196–230 (2012).
57. Zingone, A. & Oksfeldt Enevoldsen, H. The diversity of harmful algal blooms: a challenge for science and management. *Ocean & Coastal Management* 43, 725–748 (2000).
58. Van Dolah, F. M. Marine algal toxins: origins, health effects, and their increased occurrence. *Environmental Health Perspectives* 108, 133 (2000).

59. Dybas, C. L. Harmful algal blooms: biosensors provide new ways of detecting and monitoring growing threat in coastal waters. *BioScience* 53, 918–923 (2003).

60. Grant, K. S., Burbacher, T. M., Faustman, E. M. & Gratttan, L. Domoic acid: neurobehavioral consequences of exposure to a prevalent marine biotoxin. *Neurotoxicology and Teratology* 32, 132–141 (2010).

61. Graham, J. L., Loftin, K. A., Meyer, M. T. & Ziegler, A. C. Cyanotoxin mixtures and taste-and-odor compounds in cyanobacterial blooms from the Midwestern United States. *Environmental Science & Technology* 44, 7361–7368 (2010).

62. Fleming, L. E., Backer, L. C. & Baden, D. G. Overview of aerosolized Florida red tide toxins: exposures and effects. *Environmental Health Perspectives* 113, 618–620 (2005).

63. Fleming, L. E. et al. Review of Florida red tide and human health effects. *Harmful Algae* 10, 224–233 (2011).

64. Landsberg, J., Flewelling, L. & Naar, J. *Karenia brevis* red tides, brevetoxins in the food web, and impacts on natural resources: decadal advancements. *Harmful Algae* 8, 598–607 (2009).

65. Todd, E. C. Domoic acid and amnesic shellfish poisoning: a review. *Journal of Food Protection* 56, 69–83 (1993).

66. Perl, T. M. et al. An outbreak of toxic encephalopathy caused by eating mussels contaminated with domoic acid. *New England Journal of Medicine* 322, 1775–1780 (1990).

67. Bates, S. et al. Pennate diatom Nitzschia pungens as the primary source of domoic acid, a toxin in shellfish from eastern Prince Edward Island, Canada. *Canadian Journal of Fisheries and Aquatic Sciences* 46, 1203–1215 (1989).

68. Costa, L. G., Giordano, G. & Faustman, E. M. Domoic acid as a developmental neurotoxin. *Neurotoxicology* 31, 409–423 (2010).

69. Lelong, A., Hégaret, H., Soudant, P. & Bates, S. S. Pseudo-nitzschia (Bacillariophyceae) species, domoic acid and amnesic shellfish poisoning: revisiting previous paradigms. *Phycologia* 51, 168–216 (2012).

70. Lloyd, J. K., Duchin, J. S., Borchert, J., Quintana, H. F. & Robertson, A. Diarrhetic Shellfish Poisoning, Washington, USA, 2011. *Emerging Infectious Diseases* 19, 1314 (2013).

71. Watkins, S. M., Reich, A., Fleming, L. E. & Hammond, R. Neurotoxic shellfish poisoning. *Marine Drugs* 6, 431–455 (2008).

72. Etheridge, S. M. Paralytic shellfish poisoning: seafood safety and human health perspectives. *Toxicon: Official Journal of the International Society on Toxinology* 56, 108–122 (2010).

73. McGillicuddy Jr, D., Townsend, D. W., Keafer, B. A., Thomas, M. & Anderson, D. M. Georges Bank: a leaky incubator of *Alexandrium fundyense* blooms. *Deep Sea Research Part II: Topical Studies in Oceanography* 103, 163–173 (2012).

74. Faber, S. Saxitoxin and the induction of paralytic shellfish poisoning. *Journal of Young Investigators* 23, 1–7 (2012).

75. Anderson, D. M., Hoagland, P., Kaoru, Y. & White, A. W. *Estimated annual economic impacts from harmful algal blooms (HABs) in the United States.* (DTIC Document, 2000).

76. Paerl, H. W. & Scott, J. T. Throwing fuel on the fire: synergistic effects of excessive nitrogen inputs and global warming on harmful algal blooms.

Environmental Science & Technology **44**, 7756–7758, doi:10.1021/es102665e (2010).

77. Glibert, P. M. & Burkholder, J. M. Harmful algal blooms and eutrophication: "strategies" for nutrient uptake and growth outside the Redfield comfort zone. *Chinese Journal of Oceanology and Limnology* **29**, 724–738 (2011).

78. Lewitus, A. J. et al. Harmful algal blooms along the North American west coast region: history, trends, causes, and impacts. *Harmful Algae* **19**, 133–159 (2012).

79. Johnson, P. T. et al. Linking environmental nutrient enrichment and disease emergence in humans and wildlife. *Ecological Applications* **20**, 16–29 (2010).

80. Lee, C.-K., Park, T.-G., Park, Y.-T. & Lim, W. Monitoring and trends in harmful algal blooms and red tides in Korean coastal waters, with emphasis on *Cochlodinium polykrikoides*. *Harmful Algae* **30**, S3–S14 (2013).

81. Hogan, D. M. et al. Estimating the cumulative ecological effect of local scale landscape changes in South Florida. *Environmental Management* **49**, 502–515 (2012).

82. Beusen, A., Slomp, C. & Bouwman, A. Global land–ocean linkage: direct inputs of nitrogen to coastal waters via submarine groundwater discharge. *Environmental Research Letters* **8**, 034035 (2013).

83. Barnes, B. B. et al. Use of Landsat data to track historical water quality changes in Florida Keys marine environments. *Remote Sensing of Environment* **140**, 485–496 (2014).

84. Campbell, L., Henrichs, D. W., Olson, R. J. & Sosik, H. M. Continuous automated imaging-in-flow cytometry for detection and early warning of Karenia brevis blooms in the Gulf of Mexico. *Environmental Science and Pollution Research* **20**, 6896–6902 (2013).

85. Siswanto, E., Ishizaka, J., Tripathy, S. C. & Miyamura, K. Detection of harmful algal blooms of *Karenia mikimotoi* using MODIS measurements: a case study of Seto-Inland Sea, Japan. *Remote Sensing of Environment* **129**, 185–196 (2013).

86. Paerl, H. W., Hall, N. S. & Calandrino, E. S. Controlling harmful cyanobacterial blooms in a world experiencing anthropogenic and climatic-induced change. *Science of the Total Environment* **409**, 1739–1745 (2011).

87. Chen, N. et al. Nutrient enrichment and N:P ratio decline in a coastal bay–river system in southeast China: the need for a dual nutrient (N and P) management strategy. *Ocean & Coastal Management* **81**, 7–13 (2013).

88. Tomer, M., Crumpton, W., Bingner, R., Kostel, J. & James, D. Estimating nitrate load reductions from placing constructed wetlands in a HUC-12 watershed using LiDAR data. *Ecological Engineering* **56**, 69–78 (2013).

89. Glibert, P. M. et al. Vulnerability of coastal ecosystems to changes in harmful algal bloom distribution in response to climate change: projections based on model analysis. *Global Change Biology* **20**, 3845–3858 (2014).

9
Agricultural Nitrogen and Climate Change Mitigation

Given the increasing demand for nitrogen fertilisers to feed a growing human population and a potentially very large expansion in biofuel production (Chapter 11), nitrous oxide emissions from agriculture are likely to rise rapidly in coming decades[1]. The risk is that a large increase in anthropogenic nitrous oxide emissions from the agricultural sector will increasingly offset efforts to reduce carbon emissions from the energy supply and other sectors. This in turn can undermine global efforts to avoid the 2°C of post-industrial warming that is deemed to represent 'dangerous climate change'[2]. Consequently, a key mitigation challenge is to tackle nitrous oxide emissions from agriculture while still meeting the rising demands for food and biofuels around the world. Central to this battle is whether we can reduce the amount of emission from each tonne of nitrogen applied as fertiliser and each lorry-load of food produced[3].

Agriculture is the world's leading player when it comes to human-induced nitrous oxide emissions[4]. It now accounts for around 60 per cent of all anthropogenic emissions, largely due to fluxes from agricultural soils after application of nitrogen fertiliser. This not only makes the agricultural sector a major driver of climate change, but also makes it a ripe target for large-scale mitigation via the cutting of nitrous oxide emissions. How efficiently nitrogen is used – the proportion of added nitrogen that ends up in the products we consume – is a crucial determinant of global emissions. Of the roughly 100 million tonnes (1 million tonnes = 1 Tg) of nitrogen used in global agriculture each year, only about 17 per cent ends up being consumed by humans as crop, dairy or meat products[5]. Even for crop production, which is usually more efficient than livestock production, the 'nitrogen use efficiency' is generally considered to be less than 50 per cent under most on-farm conditions around the world[6]. Many options for reducing nitrous oxide emissions from agriculture therefore rely on improving this nitrogen use efficiency in one way or another.

Cropland mitigation

Fertiliser additives and application

Because microbial nitrification and denitrification are the main sources of nitrous oxide emission from agriculture, reducing how much reactive nitrogen is available to these fertiliser-interceptors is at the heart of most mitigation strategies. A well-established ploy is to apply fertiliser in combination with additives that make it more difficult for the microbes to use[7]. Slow-release fertilisers achieve this by avoiding a large burst of available nitrogen immediately after application which the crops would be unable to use up entirely. Instead, nitrogen is gradually released into the soil from pellets and the amount of excess nitrogen that the nitrifiers and denitrifiers can mop up is therefore reduced. Some fertilisers also include additives that deliberately target the microbes. These additives, called 'nitrification inhibitors', slow the activity of the nitrifiers (the microbes that convert ammonia into nitrate), allowing the crops longer access to the added nitrogen and disrupting the nitrification–denitrification supply chain that leads to nitrous oxide emissions[8,9] (Figure 9.1).

Using specific forms of fertiliser and additives can go some way to cutting emissions and boosting nitrogen access for crops, but the most effective tool for this in any farmer's fertiliser tool box is precision application. The timing of fertiliser application is crucial here, as giving the soil a huge nitrogen boost at a time when the crops are unable to make good use of it will inevitably lead to large losses and more nitrous oxide emissions. As such, a better-informed fertiliser application strategy that takes into account the changing nitrogen needs of the crop over time and matches this to supply can pay big dividends in terms of productivity, profits for the farmer and reduced emissions[10–12].

Soil and weather conditions are also important factors to consider. If the fertiliser is applied to very wet soils or during a period of heavy rain, then the chances of it being carried away in surface run-off or leaching away into groundwaters and field drains are increased[13]. Likewise, in hot and dry conditions large amounts of the nitrogen contained in fertilisers such as urea and manures may be lost to the air through volatilisation[14]. Precision application therefore entails careful matching with the nitrogen demands of the crop and adapting to short-term soil and weather conditions to minimise losses to the air and water.

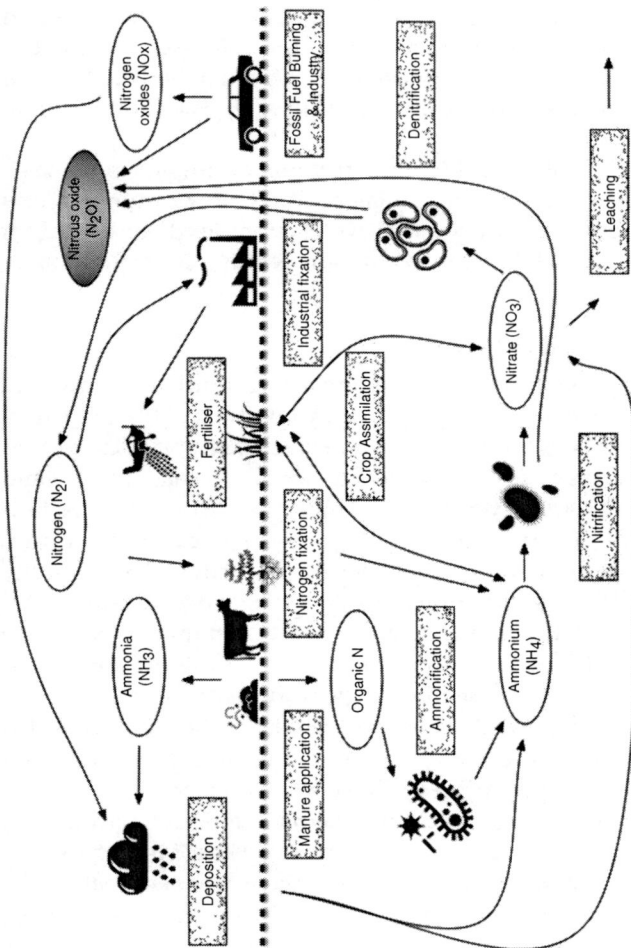

Figure 9.1   Anthropogenic terrestrial nitrogen cycle

The dotted line represents the land surface with the boxes showing the key processes by which nitrogen is cycled and the ovals showing the changing form of nitrogen as it undergoes these processes. Agriculture's impacts on the terrestrial nitrogen cycle are dominated by our direct addition of nitrogen fertilisers to soils, by nitrogen fixation by legumes, and by the redeposition of ammonia emissions produced by livestock or fertiliser application.

Source: Dave Reay

In addition to getting the amounts and timing just right, precision application can also include much more direct introduction of fertilisers to the parts of the soil where crops can use them. In the case of liquid fertilisers, injection into the soil rather than broadcasting over the soil surface can be a very effective way to reduce how much nitrogen is lost to the air via volatilisation[15]. Though such injection can mean an increase in soil nitrous oxide production in the field itself[16], the cut in volatilisation losses and the associated nitrous oxide emissions when this airborne nitrogen is redeposited can still make injection of liquid fertilisers worthwhile.

Advances in soil testing and farming technology, improved management and yield records, and development of remote sensing methods that allow farmers to identify precisely which crops need more nitrogen and which do not are all helping to put more fertiliser in the right place and at the right time[17,18].

Soil management and crop types

Improved management of soils to improve crop nutrient uptake and growth can be a 'win-win' for agriculture by boosting yields and reducing greenhouse gas emissions. Such soil management includes changes to cultivation practices designed to avoid damage to soil structure and quality. Where soils are very wet, for example, the use of heavy machinery to apply fertilisers risks not only big losses via leaching and run-off but also the compaction of soils[19]. Such compaction can drastically reduce the quality of the soil in terms of growing crops, making it difficult for roots to penetrate and providing ideal conditions for surface run-off and denitrification losses of nitrogen[20,21]. Avoiding cultivation at such times is therefore a useful strategy in the battle to improve production and cut nitrous oxide emissions. Similarly, the use of so-called 'cover crops' outside of the main growing season, whereby soil nitrogen is temporarily stored in the tissues of the cover crop plants, can serve to reduce leaching and run-off losses and maintain good soil fertility and structure[22]. These cover crops can then be ploughed back into the soil to provide a source of slow-release nitrogen for the subsequent main crop.

As the disturbance and ploughing up of soils can itself lead to big enhancements in carbon dioxide losses, a strategy of reduced or even zero tillage has been introduced in many agricultural systems for its climate change mitigation benefits[23]. In zero tillage systems the losses of soil carbon can be greatly reduced, although in some areas the reduction in soil aeration that occurs with a transition to zero tillage may help to enhance nitrous oxide production and emissions[24]. This issue is most significant for croplands where soil aeration is already limited and nitrogen inputs are large, as the absence of tillage allows increased nitrous oxide production via denitrification.

Rice production, nitrogen and methane

In rice agriculture, methane emissions due to the low-oxygen conditions induced by waterlogged soils represent a major global source of this powerful greenhouse gas. Here, nitrogen inputs can play a dual role in reducing methane emissions. Firstly, fertiliser can increase rice production in a given area, thereby reducing the amount of methane emitted per tonne of rice produced. Secondly, the added nitrogen can itself serve to reduce methane production in the low-oxygen waters and sediments by inhibiting growth and activity of the methane-producing microbes (methanogens)[25]. This reversible and relatively short-lived impact on rice paddy methane emissions must be offset by the enhancement in nitrous oxide emissions that is commonly seen with such nitrogen additions[26].

As a major source of anthropogenic methane, rice cultivation has been the focus of various climate change mitigation strategies. Many involve manipulation of water supply so as to reduce the dominance of the anoxic conditions so conducive to methane production[27]. While strategies such as more frequent draining and shorter flood periods can be very effective at cutting overall methane emissions, they may incur a nitrous oxide penalty[28]. This enhancement in emissions arises from the more frequent switching from flooded to dry soil conditions. In this case denitrification – the usual pathway for conversion of nitrate to nitrogen gas under waterlogged, low-oxygen conditions – becomes less efficient, and instead more of the nitrate is converted into nitrous oxide. In most rice paddy systems though, the climate change mitigation benefits of reduced methane emissions will far outweigh any such nitrous oxide penalty[29].

Livestock mitigation

Breeding and feeding

As in crop cultivation, reducing the amount of nitrogen input required to produce each tonne of meat or dairy product represents a keystone in efforts to reduce emissions of nitrous oxide from livestock production around the world. Selective breeding of livestock has already seen huge improvements in meat and dairy yields in many countries[30], with intensive rearing practices aimed at delivering the most cost-efficient livestock products often coinciding with more efficient nitrogen use as well. The nitrous oxide emissions caused by livestock production include those from the growing of their feed as well as the more direct emissions that arise from animal manure and urine[31]. As such, livestock products can clock up very high overall emissions across their lifecycles. Ensuring that the cultivation of animal feeds such as soybeans and silage is done in the most nitrogen-efficient way possible can therefore help to slash overall emissions from meat and dairy production.

Careful balancing of feed composition – especially protein and amino acid content – and intake with the needs of the animals at any given time can be a very effective way of reducing nitrogen wastage and the associated nitrous oxide emissions[30]. This can be difficult where livestock are grazed over a wide area and the farmer is unable to control exactly what is eaten when[32]. Regular testing of the nutrient contents of pastures and the direct feeding of dairy livestock when they are brought in for milking can provide ways for these farmers to better optimise the balance of nitrogen inputs and use.

Grazing management

Managing nitrogen inputs and cycling in grasslands is made complicated by the random additions in the form of manure and urine from grazing animals. These concentrated additions mean that even if the nitrogen balance of the field as a whole is optimised, hot spots of nitrous oxide emission may still occur[33]. Refraining from grazing livestock at very high densities for long periods or on land that is waterlogged can each help reduce nitrous oxide emissions and nitrogen losses via leaching and volatilisation. Placement and management of features such as drinking troughs, feeders and gateways can also play an important role, as areas where livestock gather frequently can quickly become compacted, manure-enriched and therefore more conducive to denitrification[34].

Regular relocation of such features can therefore help to reduce the intensity of livestock in these localised areas. More generally, the careful choice of location for grazed pasture and avoidance of livestock access to drainage channels and streams can help to avoid large losses of nitrogen to aquatic systems in the form of manure and urine[35].

Manure management

Even with well-controlled feed composition and intake, nitrogen use by livestock is relatively inefficient, with losses of 70–80 per cent in the form of manure and urine being common[36]. In intensive systems the effective collection and storage of manure can drastically reduce nitrogen losses to the air and water, with covered manure storage also helping to reduce emissions[37].

In animal housing, more frequent clearing of manure can help to address these emissions and lower the risk of large nitrogen losses via volatilisation (in hot, dry conditions) or run-off (after heavy rains)[38]. Losses of ammonia from livestock housing and manure storage via volatilisation now represent the biggest human-induced source globally[39]. Improved management of housing and storage conditions, such as frequent cleaning, and deliberate interception of ammonia before it escapes downwind from the farm (see Chapter 13) can therefore result in major reductions in just how much agricultural nitrogen ends up being deposited on unmanaged lands and waters around the world.

As with nitrogen fertilisers, application of manure to the land requires careful timing, placement and matching of inputs with plant demands. At high latitudes the winter and early spring can be an especially high-risk period for manure application. Here, large volumes of manure have often been collected from over-wintering livestock but the combination of wet conditions and slow plant growth can result in very large nitrogen losses from manure spreading.

Ruminant methane and nitrogen

Ruminant livestock, such as cattle and sheep, are a leading global source of the powerful greenhouse gas methane[40,41]. Various strategies have been developed to reduce these emissions, including altering feed quality and type – to reduce the proportion of hydrogen and carbon dioxide used by the methane-producing microbes (methanogens) in the first stomach of ruminants (the rumen) – and direct targeting of these methanogen communities using antibiotics, vaccines and alternative electron acceptors[42]. Nitrogen, usually added to feed in the form of urea or nitrate, can itself act as one of these 'alternative electron acceptors', with additions of nitrate to the feed of cattle having been shown to cut methane emissions by more than 60 per cent[43]. While nitrate and other such feed additives may give good results in terms of reduced ruminant methane, their negative impacts on animal health, productivity and nitrogen losses via manure and urine may outweigh the positives.

Agricultural nitrogen and global carbon fluxes

As discussed previously (Chapters 6 and 7), human-induced nitrogen emissions have become a powerful driver of global carbon fluxes. As agriculture is one the largest sources of reactive nitrogen emissions worldwide, and by far the biggest source of ammonia emissions, its indirect impacts on the land and ocean carbon sinks may be of major importance. Global ammonia emissions from agriculture are estimated to be over 30 million tonnes of nitrogen per year[39] and are concentrated in Southern Asia, Northern Europe and North America[44]. The bulk of these emissions are redeposited on land, with around 40 per cent being deposited in coastal areas and the open ocean. As most agricultural land already receives large inputs of nitrogen as fertiliser, additional inputs from ammonia deposition are unlikely to have a major effect on carbon sequestration in the soils. However, in unmanaged areas such as peatlands, increasing deposition could potentially increase soil carbon stocks, but so far there are not enough data available to know how important this response is. Likewise, more agricultural ammonia deposition to freshwater ecosystems, coastal areas and oceans may promote carbon dioxide uptake, but its true extent remains poorly quantified.

Rather better understood than soil and aquatic carbon sinks is the effect of increasing nitrogen deposition on the global forest sink. Current ammonia emissions from agriculture result in an estimated enhancement of up to one billion tonnes (1 Pg) per year in the global forest carbon sink, assuming the 60 per cent of ammonia deposited to land is evenly dispersed and that 200 grams of extra carbon are stored for every gram of new nitrogen deposited on the forests[44]. In reality, the dispersal of agricultural ammonia is far from even. Much is deposited within the same region from which it was emitted rather than being spread right around the planet, and the average response of the forest carbon sink is probably much lower than the 200 grams per gram super-boost assumed here. Any net climate-forcing benefits of these agricultural ammonia losses are further eroded by the enhancement of soil and water nitrous oxide emissions. If just one per cent of ammonia is then emitted as nitrous oxide, it would equate to an extra 300,000 tonnes of nitrous oxide-nitrogen being released – a 'climate mitigation penalty' equivalent to 140 million tonnes of carbon dioxide.

Ammonia emissions from agriculture, and therefore deposition, are expected to grow even more as global demand for meat and dairy products increases. Over the next few decades emissions from agriculture are expected to double in regions such as Southeast Asia and Central America[45]. In theory, this ongoing surge in agricultural nitrogen emissions could help further enhance carbon sinks thereby mitigating climate change. However, the concentration of this extra deposition in areas already experiencing high nitrogen loads is likely to mean that the carbon sink benefits are small and far outweighed by the negative impacts to ecosystems and human health (Chapter 6).

Finally, the production of nitrogen fertilisers for agriculture is itself an energy-intensive process, requiring large amounts of fossil fuel burning (usually in the form of natural gas or coal) and producing significant volumes of nitrogen oxide (NOx) gases and nitrous oxide along with the fossil fuel carbon dioxide. The bulk of carbon dioxide emissions arise from the Haber–Bosch process, whereby ammonia is formed by combining nitrogen and hydrogen gases at high temperature (Chapter 2). Natural gas (CH_4) is first used to obtain the hydrogen feedstock – accounting for about two-thirds of the total fossil fuel use – after which more gas or coal is burned to create the heat that drives the chemical reaction of nitrogen with hydrogen[46]. Some of the carbon dioxide produced is captured and reused to make urea and other industrial products, but the bulk generated via fertiliser production is emitted into the atmosphere.

Around 80 per cent of all ammonia produced in this way is destined for agricultural use, with some being further converted into nitric acid to produce nitrate-based fertilisers[47]. Globally, production of nitrogen fertilisers is thought to be responsible for the equivalent of between 350 million and 700 million tonnes of carbon dioxide each year via production, distribution and storage[48] – which is still much less than that emitted from soils in the form of nitrous oxide after these fertilisers are applied.

While nitrous oxide emissions from the industrial production of nitric acid have been radically reduced in most nations (Chapter 4), carbon dioxide emissions from fertiliser production have proved more difficult to address. A widespread switch from coal to gas as a fossil fuel source has helped to reduce the carbon intensity of production, but the overall volume of production has continued to increase and therefore absolute emissions have remained high. The choice of fossil fuel for ammonia production reflects market prices and, if coal prices are low, a switch back towards use of this more carbon-intensive energy source may occur. Similarly, substitution of gas or coal with renewable energy sources for ammonia production could provide huge reductions in the carbon intensity of fertilisers, but also incur extra costs for the manufacturers. Ultimately, the most far-reaching way of cutting these production-side greenhouse gas emissions is to reduce the total amount of fertiliser required by agriculture in the first place. Improvements in nitrogen use efficiency can certainly help in this goal, although meeting the food and energy needs of a projected human population of almost 10 billion by 2050[49] will inevitably require much more fertiliser, rather than much less.

Optimising global agriculture

Using strategies such as slow-release fertilisers, precision application and optimised timing it is estimated that nitrogen use efficiency could be increased by up to 50 per cent in much of the world[50–52]. In some under-fertilised regions, including large areas of Africa[53], much more fertiliser nitrogen may be needed to increase yields. So, whilst the nitrous emissions in these areas might be expected to increase, the overall emissions per unit of agricultural product may be significantly decreased.

Underpinning these and the various other interventions to nitrogen use in crop or livestock production described so far in this chapter – and often dwarfing them in terms of potential climate change mitigation benefits – is the goal of optimising agriculture on a global scale by improving efficiency of production at a local level. For instance, the types of crops grown in a given area can play a major role in the overall efficiency of nitrogen use and the resulting greenhouse gas fluxes. Where soil conditions are marginal for crop growth due to soil structure, chemistry and climate, addition of large amounts of fertilisers may be required per tonne of crop produced, with nitrogen losses then being increased. Likewise, access to fertilisers, pesticides, cultivation technology and expertise in areas ideal for cultivation of specific crops may be limited. Trade barriers, farming subsidies, consumer demand and wide fluctuations in market prices[54] can all lead to the cultivation of crops on land that is far from ideal, while prime land sits unused elsewhere.

In livestock and dairy production, similar cross-border barriers to improved nitrogen use efficiency apply, with variable access to resources such as veterinary support, animal housing, transport and manure storage all meaning that opportunities to increase efficiency and reduce greenhouse gas emissions are lost. No matter how 'low-nitrogen' the initial production of a crop or herd of cattle is, if the whole crop or herd is then lost due to pests or disease the overall nitrogen use efficiency is still zero. Ultimately, the global mitigation of nitrogen wastage and greenhouse gas emissions in agriculture will require inequalities such as fertiliser availability and artificially high and low food prices to be addressed.

References

1. Reay, D. S. et al. Global agriculture and nitrous oxide emissions. *Nature Climate Change* 2, 410–416, doi:10.1038/nclimate1458 (2012).
2. Anderson, K. & Bows, A. Beyond 'dangerous' climate change: emission scenarios for a new world. *Philosophical Transactions of the Royal Society A: Mathematical, Physical and Engineering Sciences* 369, 20–44 (2011).
3. Sutton, M. A. et al. *The European nitrogen assessment: sources, effects and policy perspectives.* (Cambridge University Press, 2011).
4. Mosier, A. et al. Closing the global N(2)O budget: nitrous oxide emissions through the agricultural nitrogen cycle – OECD/IPCC/IEA phase II development of IPCC guidelines for national greenhouse gas inventory methodology. *Nutrient Cycling in Agroecosystems* 52, 225–248, doi:10.1023/a:1009740530221 (1998).

5. Braun, E. *Reactive nitrogen in the environment: too much or too little of a good thing.* (UNEP/Earthprint, 2007).
6. Tilman, D., Cassman, K. G., Matson, P. A., Naylor, R. & Polasky, S. Agricultural sustainability and intensive production practices. *Nature* **418**, 671–677 (2002).
7. Smith, K., McTaggart, I. & Tsuruta, H. Emissions of N2O and NO associated with nitrogen fertilization in intensive agriculture, and the potential for mitigation. *Soil Use and Management* **13**, 296–304 (1997).
8. Mosier, A., Duxbury, J., Freney, J., Heinemeyer, O. & Minami, K. Nitrous oxide emissions from agricultural fields: assessment, measurement and mitigation. *Plant and Soil* **181**, 95–108 (1996).
9. Di, H., Cameron, K. & Sherlock, R. Comparison of the effectiveness of a nitrification inhibitor, dicyandiamide, in reducing nitrous oxide emissions in four different soils under different climatic and management conditions. *Soil Use and Management* **23**, 1–9 (2007).
10. Robertson, G. P. & Vitousek, P. M. Nitrogen in agriculture: balancing the cost of an essential resource. *Annual Review of Environment and Resources* **34**, 97–125 (2009).
11. Wallace, A. High-precision agriculture is an excellent tool for conservation of natural resources. *Communications in Soil Science & Plant Analysis* **25**, 45–49 (1994).
12. Ladha, J. K., Pathak, H., Krupnik, T. J., Six, J. & van Kessel, C. Efficiency of fertilizer nitrogen in cereal production: retrospects and prospects. *Advances in Agronomy* **87**, 85–156 (2005).
13. Di, H. & Cameron, K. Nitrate leaching in temperate agroecosystems: sources, factors and mitigating strategies. *Nutrient Cycling in Agroecosystems* **64**, 237–256 (2002).
14. McGarry, S., O'Toole, P. & Morgan, M. Effects of soil temperature and moisture content on ammonia volatilization from urea-treated pasture and tillage soils. *Irish Journal of Agricultural Research* **26**, 173–182 (1987).
15. Sommer, S. G. & Hutchings, N. Ammonia emission from field applied manure and its reduction—invited paper. *European Journal of Agronomy* **15**, 1–15 (2001).
16. Flessa, H. & Beese, F. Laboratory estimates of trace gas emissions following surface application and injection of cattle slurry. *Journal of Environmental Quality* **29**, 262–268 (2000).
17. Bausch, W. C. & Diker, K. Innovative remote sensing techniques to increase nitrogen use efficiency of corn. *Communications in Soil Science and Plant Analysis* **32**, 1371–1390 (2001).
18. Meisinger, J. & Delgado, J. Principles for managing nitrogen leaching. *Journal of Soil and Water Conservation* **57**, 485–498 (2002).
19. Chamen, T. et al. Prevention strategies for field traffic-induced subsoil compaction: a review: Part 2. Equipment and field practices. *Soil and Tillage Research* **73**, 161–174 (2003).
20. Lipiec, J. & Stepniewski, W. Effects of soil compaction and tillage systems on uptake and losses of nutrients. *Soil and Tillage Research* **35**, 37–52 (1995).
21. Barken, L., Bøsrresen, T. & Njøss, A. Effect of soil compaction by tractor traffic on soil structure, denitrification, and yield of wheat (*Triticum aestivum* L.). *Journal of Soil Science* **38**, 541–552 (1987).

22. Shipley, P., Messinger, J. & Decker, A. Conserving residual corn fertilizer nitrogen with winter cover crops. *Agronomy Journal* **84**, 869–876 (1992).
23. Smith, P., Powlson, D. S., Glendining, M. J. & Smith, J. U. Preliminary estimates of the potential for carbon mitigation in European soils through no-till farming. *Global Change Biology* **4**, 679–685 (1998).
24. Rochette, P., Angers, D. A., Chantigny, M. H. & Bertrand, N. Nitrous oxide emissions respond differently to no-till in a loam and a heavy clay soil. *Soil Science Society of America Journal* **72**, 1363–1369 (2008).
25. Klüber, H. D. & Conrad, R. Effects of nitrate, nitrite, NO and N2O on methanogenesis and other redox processes in anoxic rice field soil. *FEMS Microbiology Ecology* **25**, 301–318 (1998).
26. Cai, Z. et al. Methane and nitrous oxide emissions from rice paddy fields as affected by nitrogen fertilisers and water management. *Plant and Soil* **196**, 7–14 (1997).
27. Ratering, S. & Conrad, R. Effects of short-term drainage and aeration on the production of methane in submerged rice soil. *Global Change Biology* **4**, 397–407 (1998).
28. Towprayoon, S., Smakgahn, K. & Poonkaew, S. Mitigation of methane and nitrous oxide emissions from drained irrigated rice fields. *Chemosphere* **59**, 1547–1556 (2005).
29. Zou, J., Huang, Y., Jiang, J., Zheng, X. & Sass, R. L. A 3-year field measurement of methane and nitrous oxide emissions from rice paddies in China: effects of water regime, crop residue, and fertilizer application. *Global Biogeochemical Cycles* **19**, GB2021 (2005).
30. Eckard, R., Grainger, C. & De Klein, C. Options for the abatement of methane and nitrous oxide from ruminant production: a review. *Livestock Science* **130**, 47–56 (2010).
31. Beauchemin, K. A., Henry Janzen, H., Little, S. M., McAllister, T. A. & McGinn, S. M. Life cycle assessment of greenhouse gas emissions from beef production in western Canada: a case study. *Agricultural Systems* **103**, 371–379 (2010).
32. De Klein, C. & Eckard, R. Targeted technologies for nitrous oxide abatement from animal agriculture. *Animal Production Science* **48**, 14–20 (2008).
33. Flessa, H., Dörsch, P., Beese, F., König, H. & Bouwman, A. Influence of cattle wastes on nitrous oxide and methane fluxes in pasture land. *Journal of Environmental Quality* **25**, 1366–1370 (1996).
34. Šimek, M., Brůček, P., Hynšt, J., Uhlířová, E. & Petersen, S. O. Effects of excretal returns and soil compaction on nitrous oxide emissions from a cattle overwintering area. *Agriculture, Ecosystems & Environment* **112**, 186–191 (2006).
35. Line, D., Harman, W., Jennings, G., Thompson, E. & Osmond, D. Nonpoint-source pollutant load reductions associated with livestock exclusion. *Journal of Environmental Quality* **29**, 1882–1890 (2000).
36. Rotz, C. Management to reduce nitrogen losses in animal production. *Journal of Animal Science* **82**, E119–E137 (2004).
37. Chadwick, D. Emissions of ammonia, nitrous oxide and methane from cattle manure heaps: effect of compaction and covering. *Atmospheric Environment* **39**, 787–799 (2005).
38. Monteny, G. & Erisman, J. Ammonia emission from dairy cow buildings: a review of measurement techniques, influencing factors and possibilities for reduction. *NJAS Wageningen Journal of Life Sciences* **46**, 225–247 (1998).

39. Beusen, A., Bouwman, A., Heuberger, P., Van Drecht, G. & Van Der Hoek, K. Bottom-up uncertainty estimates of global ammonia emissions from global agricultural production systems. *Atmospheric Environment* 42, 6067–6077 (2008).

40. Stocker, T. *Climate change 2013: the physical science basis: Working Group I contribution to the Fifth assessment report of the Intergovernmental Panel on Climate Change.* (Cambridge University Press, 2014).

41. Kelliher, F. & Clark, H. Ruminants. In *Methane and Climate Change*, edited by D. Reay, P. Smith & A. van Amstel, 136–150 (Earthscan, London, 2010).

42. Martin, C., Morgavi, D. & Doreau, M. Methane mitigation in ruminants: from microbe to the farm scale. *Animal: An International Journal of Animal Bioscience* 4, 351–365 (2010).

43. Nolan, J., Hegarty, R., Hegarty, J., Godwin, I. & Woodgate, R. Effects of dietary nitrate on fermentation, methane production and digesta kinetics in sheep. *Animal Production Science* 50, 801–806 (2010).

44. Reay, D. S., Dentener, F., Smith, P., Grace, J. & Feely, R. A. Global nitrogen deposition and carbon sinks. *Nature Geoscience* 1, 430–437, doi:10.1038/ngeo230 (2008).

45. Dentener, F. et al. Nitrogen and sulfur deposition on regional and global scales: a multimodel evaluation. *Global Biogeochemical Cycles* 20, 1–21 (2006).

46. Wood, S. & Cowie, A. A review of greenhouse gas emission factors for fertiliser production. *IEA Bioenergy Task* 38, 20 pp. (Research and Development Division, State Forests of New South Wales, Australia, 2004).

47. Wiesen, P. Abiotic nitrous oxide sources: chemical industry and mobile and stationary combustion systems. In *Nitrous Oxide and Climate Change*, edited by K. Smith, 190–209 (Earthscan, London, 2010).

48. Flynn, H. C. & Smith, P. Greenhouse gas budgets of crop production–current and likely future trends. *France. (PDF can be downloaded at www.sustainablecropnutrition.com, file name: 2010_ifa_greenhouse_gas.pdf)* (2010).

49. Cohen, J. E. Human population: the next half century. *Science* 302, 1172–1175 (2003).

50. Del Grosso, S. J. & Grant, D. W. Reducing agricultural GHG emissions: role of biotechnology, organic systems and consumer behavior. *Carbon Management* 2, 505–508 (2011).

51. Smil, V. *Enriching the earth: Fritz Haber, Carl Bosch, and the transformation of world food production.* (MIT Press, 2001).

52. Erisman, J. W., Sutton, M. A., Galloway, J., Klimont, Z. & Winiwarter, W. How a century of ammonia synthesis changed the world. *Nature Geoscience* 1, 636–639, doi:10.1038/ngeo325 (2008).

53. Sanchez, P. A. Soil fertility and hunger in Africa. *Science (Washington)* 295, 2019–2020 (2002).

54. Mayrand, K., Dionne, S., Paquin, M. & Pageot-LeBel, I. The economic and environmental impacts of agricultural subsidies: an assessment of the 2002 US Farm Bill & Doha Round. *Unisféra International Centre* (2003).

10
Nitrogen in Food and Climate Change Mitigation

The revolution in global food production that nitrogen fertilisers and improved agricultural practices have delivered in the last half century has at least reduced the proportion of hungry or undernourished people in the world – down from a third in the 1960s to less than a quarter today. But population growth and stark inequalities in food supply between rich and poor nations have meant that the absolute number has changed little. Over 800 million people are still classed as undernourished[1,2], while each year, in the UK and US alone, a mass of food equivalent to the needs of some 80 million people is simply thrown away. Worldwide, each person now consumes about half a tonne of food annually, most of this as cereals, fruit and vegetables, with about a quarter in the form of meat, seafood and dairy produce[3]. Hidden within this global average are national diets that are driving obesity in their own populations while stripping calories from the already-inadequate food supplies of others. Meat consumption is at the enlarged heart of this global see-saw of feast and famine and, with the high nitrogen intensity of its production[4], presents a host of risks and opportunities in the battle to avoid dangerous climate change.

Dietary choice and nitrogen wastage

As human populations become wealthier, the demand for a more varied and meat-intensive diet tends to grow[2]. In the past four decades the global production of pork, beef, lamb and chicken has tripled[3] – there are now two chickens for every person on the planet. In the US the average person consumes close to their own body weight in meat every year[5]. Huge amounts of grain are diverted from direct human consumption to feed the intensively reared hoards of soon-to-be burgers, chops and drumsticks[4], with meat requiring up to seven times more land than crops of the same food value. This trend has big implications for the use, and waste, of nitrogen. For instance, by taking the more direct pathway to the plates of vegetarians, one-sixth of the fertiliser used during crop production will actually make it into the nut roasts and aubergine bakes. For meat-loving omnivores this proportion plummets to less than one-twentieth[6].

In addition to measures that directly reduce nitrous oxide emissions in production (Chapter 9), there exists significant potential for climate change mitigation via the demand side through changes to human dietary choice[7]. Just as a shift towards a greater per capita caloric intake and increased proportion of animal products in diets is expected to enhance agricultural nitrous oxide emissions, policies that achieve a reduction in animal product consumption[8,9], or that successfully address excessive caloric intake[10], can reduce them. For example, a recent projection of greenhouse gas emissions from global agricultural up to the middle of the 21st century estimated that if current diets remained the same, emissions would rise to the equivalent of over eight billion tonnes of carbon dioxide each year – an increase of 60 per cent compared to 1995[9]. If average human diets over this period actually became even more meat-intensive than today[11], emissions would more than double.

However, this study also examined a future where diets became less meat-intensive than today. In that future scenario of decreased meat consumption, global soil nitrous oxide emissions fell almost one-quarter by the middle of the century and overall agricultural greenhouse gas emissions decreased[9]. Effectively, this scenario allows the amount of food produced to increase so as to meet the demands of a growing population, but at the same time delivers a large cut in greenhouse gas emissions.

Turning Japanese

Such an impressive potential for cutting agricultural nitrous oxide emissions via dietary choice can also been seen where cultures already achieving high food availability alongside low amounts of meat intake are taken as a global benchmark. Japan is a good example of this, with an average diet that has a high proportion of fish instead of animal meats like chicken. As chicken meat production is nitrogen-intensive and often leads to high nitrous oxide emissions per kilogram produced, a global transition in chicken meat consumption to the low levels seen in Japan could yield impressive overall reductions in emissions[11] (Box 10.1).

Overall, dietary change that results in reduced consumption of food types with low nitrogen efficiencies in production, such as meat and dairy,

Box 10.1

Global convergence on a Japanese diet

By combining average per capita poultry meat intake in the developed and developing world with projected population change up to 2020, and by then applying an estimate of nitrous oxide emissions in the 'production phase' of poultry meat, global emissions would increase by around 20 per cent. However, if per capita intake in the rest of the developed world over this period were to instead converge with the relatively low levels estimated for Japan, global poultry meat-related nitrous oxide emissions would actually decrease by 20 per cent – a complete reversal of the current trend.

Large potential reductions are also seen when pig and sheep meat consumption are examined. Here a global convergence towards a Japanese-style diet would deliver estimated cuts of 10–15 per cent in nitrous oxide emissions from production of these meats by 2020[11]. Clearly, such estimates provide only an indication of how cuts in agricultural emissions may be achieved through dietary change. Any apparent reduction observed with a switch from poultry, pig or sheep meat consumption must be set against the resultant increases in consumption of other foodstuffs – an enormous increase in seafood in the case of global convergence with the Japanese diet. Consequently, a serious gap in current scientific knowledge is exactly how much nitrous oxide is emitted by fast-emerging food sectors like aquaculture. This industry has expanded very quickly over the last few decades, but the amount of nitrous oxide it produces globally remains poorly quantified[12]. If emissions per kilogram of seafood actually end up being similar to those for the production of chicken, pig and sheep meat, then the climate change benefits of a Japanese-style diet would largely disappear.

may have potential in addressing global climate change. But the reality of dietary change in much of the world is that it is on a more, not less, meat-intensive trajectory[2]. Changing this pathway will require a host of interventions ranging from altered price incentives and subsidies to the use of public health and environmental education campaigns[13]. If the rising global demand for meat and dairy products cannot be eased by a move to less meat-intensive diets, then the ongoing efforts to cut total nitrous oxide emissions in livestock agriculture risk being badly undermined.

Excess calorie intake and nitrogen

Along with changing the composition of average diets to those that are less nitrogen-intensive, there is also the potential for reducing agricultural demand (and so emissions) by targeting excess consumption. Globally, more than one billion people are now classed as 'overweight' or 'obese', with major negative consequences for human health, such as Type 2 diabetes and increased risk of ischaemic heart disease[8,14]. The excessive calorie intake common to the diets of these individuals itself represents a significant inefficiency in terms of global food nitrogen use.

Box 10.2

Chicken dinners and nitrous oxide in the US

Between 1970 and today obesity rates in the adult US population rose from around 14 per cent to more than 35 per cent. Over this time per capita intake of chicken also rose rapidly – from 19 kg per year in 1970 to a hefty 46 kg in 2014. Under a 'business-as-usual' scenario the average US citizen will be consuming almost 50 kg of chicken meat a year by 2020[2,3].

Reducing obesity rates to their 1970s levels would have to involve calorie-intake reductions across a wide range of foods but, assuming that chicken meat consumption was part of this overall change, a return to 1970s levels is conceivable. If this were to happen, then overall chicken meat consumption in the US each year would fall from 14.5 million tonnes to just 6 million tonnes. In addition to the health benefits that may accrue, this huge drop in chicken meat demand could radically cut environment-damaging nitrogen losses during production.

Every tonne of chicken meat produced results in around 30 kg of nitrogen loss in the form of nitrate, 40 kg as ammonia and 6.3 kg as nitrous oxide[15]. Under the obesity-lowering scenario of a return to the 1970s levels of chicken consumption, losses of nitrate via leaching and run-off could therefore be reduced by 255,000 tonnes of nitrogen, losses of ammonia to the air by 340,000 tonnes and emissions of the powerful greenhouse gas nitrous oxide by over 53,000 tonnes of nitrogen per year – equivalent to a cut of 25 million tonnes of carbon dioxide or taking 5 million cars off the road. As such, the human health and environmental 'win-win' of addressing excess calorie intake also represents a major opportunity to reduce nitrogen wastage and halt the rise in nitrogen-induced global climate change.

The dietary composition of these overweight and obese populations also tends towards high nitrogen intensity, with a key factor being the excess animal protein intake in the form of meat and dairy products[8]. By avoiding excess calorie intake and, in particular, by doing this through reduced meat and dairy intake, the upstream ripples to global livestock production and greenhouse gas emissions could be enormous (Box 10.2).

Food waste and nitrogen

Alongside interventions aimed at reducing the environmental impact of human diets, reductions in food loss and wastage may also be a big help in addressing agriculture's expanding climate change footprint. Food loss and wastage – the food that is produced for human consumption but that never makes it to our mouths – accounts for more than 30 per cent of all food production in the world each year[16]. This enormous wastage of food occurs for a wide range of reasons. In the developed world food wastage tends to be most common in the 'consumer phase', with households throwing away around one-third of the food they buy because it has gone past its use-by date or the servings are simply too large[17]. Much of this food waste ends up in landfill sites where it can incur a climate change penalty in terms of methane and nitrous oxide emissions as the waste decomposes[18]. Much more important though, in terms of the climate and environmental impacts of this food wastage, are the implications it has for emissions in the 'production phase' (i.e. back on the farm).

Every tonne of food that is wasted means another tonne of food that needs to be produced, and with this extra production inevitably comes more nitrogen use and more nitrous oxide emissions. As some foods, such as meat and dairy products, are especially nitrogen-intensive, their wastage can incur very large penalties in terms of enhanced nitrous oxide emissions from the agricultural sector. For instance, in Europe the 'nitrogen footprint' of beef production is estimated to be some 500 grams of reactive nitrogen per kilogram, while the footprint for production of most vegetables is only about 2 grams nitrogen per kilogram[19]. As such, interventions in the food supply chain that can reduce consumer wastage of high-nitrogen foods like meat – improved labelling, for instance – can generate an upstream ripple that radically reduces nitrogen use and losses during production.

In much of the developing world consumer-phase wastage rates are actually relatively low, but overall food loss and wastage rates are still around the 30–40 per cent level due to the damage to food that occurs between the farm and the intended consumers. Key reasons for these losses include a lack of storage for harvested foods, absence of adequate packaging and refrigeration, and long transport times to get food from farm to market due to poor transport infrastructure[16]. In a world where 800 million people are classed as undernourished and where agricultural emissions are playing an ever more important role in driving climate change, the potential benefits of reducing food loss and wastage have become a very active area of interest for researchers and policy makers.

Box 10.3

Milk waste avoidance and nitrous oxide

Of the 13 million tonnes of raw milk produced for domestic consumption in the UK in 2009, some 360,000 tonnes (~3 per cent) was wasted by consumers[11]. The primary reasons for this wastage were too much milk being purchased, much of which had then gone beyond its use-by date, and too much milk being served, the unwanted portion of which was then being poured away. Of the 360,000 tonnes of milk that was bought but went unconsumed by UK households in 2009, more than 99 per cent was designated as 'avoidable wastage'[17].

Because dairy cattle require large feed inputs and as the conversion of nitrogen in these feeds is rather inefficient, every 10,000 litres of milk production in the UK is estimated to result in 7 kg of nitrogen losses as nitrous oxide[15]. For the UK as a whole then, cutting out all avoidable milk wastage in the consumer phase would mean an upstream reduction in dairy farming nitrous oxide emissions of an impressive 250 tonnes of nitrogen per year.

Although milk is a relatively nitrous oxide-intensive product and constitutes a large proportion of avoidable food waste[16], the same kind of far-reaching benefits to climate change mitigation in the agricultural sector can be seen in avoiding consumer-phase wastage of other major foods[11].

By tackling the avoidable consumer-phase wastage of milk, poultry meat, pig meat, sheep meat and potatoes, nitrous oxide emission reductions in the UK totalling more than two thousand tonnes of nitrogen each year seem achievable – a major contribution to the nation's climate change mitigation efforts and to increasing overall efficiency in nitrogen use in the agricultural sector.

Food nitrogen and climate change mitigation

Given the magnitude of greenhouse gas emissions from the agricultural sector, strategies that reduce food loss and wastage clearly have great potential as effective tools in global climate change mitigation. Looking at the global food loss and wastage rate of around 30 per cent and the total amount of nitrous oxide that is emitted by agriculture each year – which is over three million tonnes of nitrogen per year – complete avoidance of food wastage could possibly result in a cut in global nitrous oxide emissions of around one million tonnes of nitrogen per year. This is equivalent to emissions savings of almost half a billion tonnes of carbon dioxide a year (due to nitrous oxide being such a powerful greenhouse gas) or of taking around 100 million cars off the world's roads. The true potential for such mitigation will inevitably vary depending on food type, production and location (Box 10.3). Some food loss and wastage will always occur, such as the inedible components of meat products like bones.

Achieving large cuts in avoidable food wastage requires interventions focused on altering consumer behaviour. These may include the encouragement of smaller, more frequent food shopping trips and recommendations on reduced portion sizes. Public and school education initiatives can be effective strategies for delivery of this information, as also can collaboration of government with food retailers on issues such as 'buy one, get one free' promotions that can play a role in over-consumption and wastage[20].

The huge potential for cutting particular types of food wastage and the associated nitrous oxide emissions can also be estimated at the global scale, though the data on amounts of waste and the nation-specific emissions from food production are not always as complete as they are for the UK. Using the global average for food supply-chain loss (~30 per cent) and combining how much of five key food types (milk, potatoes, poultry, pig and sheep meat) is produced each year with the UK figures for their nitrous oxide footprint gives loss- and wastage-associated emissions of more than 200,000 tonnes of nitrogen a year. This loss and wastage for these five food types alone constitutes ~3 per cent of worldwide nitrous oxide emissions from agriculture. Again, the proportion that is realistically 'avoidable' will vary greatly depending on the food type, location and stage in the supply chain, but enormous reductions in emissions by tackling food loss and wastage in the production, distribution and the consumer phases do seem possible[11].

Climate change impacts and food waste

Food 'loss and wastage' is usually defined as the mass of a food directed for human consumption that is lost or wasted in the supply chain[16]. Food 'losses' refer to a decrease in the edible food mass in the production, post-harvest and processing phases, while food 'wastage' tends to refer to a decrease in the edible food mass in the retail and consumer phases. As mentioned previously, in much of the developed world it is in the consumer phase that the bulk of food loss occurs, while in many developing nations it is losses in the production and pre-consumer phases that dominate.

As climate change intensifies, its negative impacts on agricultural production and the overall food supply chain may also increase around the world[21-23]. Changes in precipitation and more intense or frequent severe weather events, such as droughts and floods, are projected to cause more and more damage to crop and livestock production. Higher temperatures may also reduce productivity, increasing heat stress in livestock[24] and damaging the flowering and germination rates of many crop types[25]. Pest, weed and disease threats to agriculture may increase in some regions[26], with higher temperatures also threatening to enhance low-level ozone formation and damage to crop growth[27].

Post-harvest, climate change may further increase food loss and wastage rates through higher temperatures which, in turn, increase the growth rate of spoilage organisms, such as bacteria and fungi, and therefore the speed at which harvested food deteriorates[28]. This faster deterioration reduces the time available in which to get the food from farm to consumer, needing faster farm-to-market transport or requiring more refrigeration capacity and the electricity to power it[29]. Flooding and severe weather events may also damage food storage and processing facilities or the power infrastructure they require for cooling and drying. The food transport system itself may be negatively affected by intensifying climate change, with transport times greatly increased by the loss of road and rail links due to flooding and storms. Finally, higher temperatures pose a challenge for food longevity in the consumer phase in all nations, with the acceleration of spoilage reducing the shelf lives of many perishable products and threatening a marked increase, rather than a decrease, in consumer-phase food wastage. Successfully reducing food loss and wastage around the world will therefore require an increase in the resilience of both food production and supply to future climate change[30].

References

1. Barrett, C. B. Measuring food insecurity. *Science* **327**, 825–828 (2010).
2. Alexandratos, N. & Bruinsma, J. World agriculture towards 2030/2050: the 2012 revision. (ESA Working paper Rome, FAO, 2012).
3. FAOSTAT. Statistical databases. *Food and Agriculture Organization of the United Nations* (http://faostat.fao.org) (2009).
4. Bouwman, A. & Booij, H. Global use and trade of feedstuffs and consequences for the nitrogen cycle. *Nutrient Cycling in Agroecosystems* **52**, 261–267 (1998).
5. Daniel, C. R., Cross, A. J., Koebnick, C. & Sinha, R. Trends in meat consumption in the USA. *Public Health Nutrition* **14**, 575–583 (2011).
6. Bleken, M. A. & Bakken, L. R. The nitrogen cost of food production: Norwegian society. *Ambio* **26**, 134–142 (1997).
7. Stehfest, E. et al. Climate benefits of changing diet. *Climatic Change* **95**, 83–102 (2009).
8. McMichael, A. J., Powles, J. W., Butler, C. D. & Uauy, R. Food, livestock production, energy, climate change, and health. *The Lancet* **370**, 1253–1263 (2007).
9. Popp, A., Lotze-Campen, H. & Bodirsky, B. Food consumption, diet shifts and associated non-CO_2 greenhouse gases from agricultural production. *Global Environmental Change* **20**, 451–462 (2010).
10. Edwards, P. & Roberts, I. Population adiposity and climate change. *International Journal of Epidemiology* **38**, 1137–1140 (2009).
11. Reay, D. S. et al. Global agriculture and nitrous oxide emissions. *Nature Climate Change* **2**, 410–416, doi:10.1038/nclimate1458 (2012).
12. Williams, J. & Crutzen, P. Nitrous oxide from aquaculture. *Nature Geoscience* **3**, 143–143 (2010).
13. Reay, D. S. et al. Societal choice and communicating the European nitrogen challenge. In *The European Nitrogen Assessment*, edited by M. Sutton et al., 585–601 (Cambridge University Press, UK, 2011).
14. Kelly, T., Yang, W., Chen, C., Reynolds, K. & He, J. Global burden of obesity in 2005 and projections to 2030. *International Journal of Obesity* **32**, 1431–1437 (2008).
15. Williams, A., Audsley, E. & Sandars, D. Determining the environmental burdens and resource use in the production of agricultural and horticultural commodities. Main report. Defra Research Project IS0205. Cranfield University and Defra, Bedford. *There is no corresponding record for this reference* (2012).
16. Gustavsson, J., Cederberg, C., Sonesson, U., Van Otterdijk, R. & Meybeck, A. *Global food losses and food waste: extent, causes and prevention.* (FAO, Rome, 2011).
17. Quested, T. & Johnson, H. *Household food and drink waste in the UK: final report.* (Wastes & Resources Action Programme [WRAP], 2009).
18. Adhikari, B. K., Barrington, S. & Martinez, J. Predicted growth of world urban food waste and methane production. *Waste Management & Research* **24**, 421–433 (2006).
19. Leip, A., Weiss, F., Lesschen, J. & Westhoek, H. The nitrogen footprint of food products in the European Union. *The Journal of Agricultural Science* **152**, 1–14 (2013).

20. Gunders, D. Wasted: how America is losing up to 40 percent of its food from farm to fork to landfill. *Natural Resources Defense Council Issue Paper. August. This report was made possible through the generous support of The California Endowment* (2012).
21. Parry, M. L., Rosenzweig, C., Iglesias, A., Livermore, M. & Fischer, G. Effects of climate change on global food production under SRES emissions and socio-economic scenarios. *Global Environmental Change* **14**, 53–67 (2004).
22. Schmidhuber, J. & Tubiello, F. N. Global food security under climate change. *Proceedings of the National Academy of Sciences of the United States of America* **104**, 19703–19708 (2007).
23. Liverman, D. & Kapadia, K. Food systems and the global environment: an overview. In *Food Security and Global Environmental Change*, edited by J. Ingram, P. Ericksen & D. Liverman, 3–24 (Earthscan, 2010).
24. Thornton, P., Van de Steeg, J., Notenbaert, A. & Herrero, M. The impacts of climate change on livestock and livestock systems in developing countries: a review of what we know and what we need to know. *Agricultural Systems* **101**, 113–127 (2009).
25. Schlenker, W. & Roberts, M. J. Nonlinear temperature effects indicate severe damages to US crop yields under climate change. *Proceedings of the National Academy of Sciences of the United States of America* **106**, 15594–15598 (2009).
26. Rosenzweig, C., Iglesias, A., Yang, X., Epstein, P. R. & Chivian, E. Climate change and extreme weather events; implications for food production, plant diseases, and pests. *Global Change & Human Health* **2**, 90–104 (2001).
27. Reilly, J. et al. Global economic effects of changes in crops, pasture, and forests due to changing climate, carbon dioxide, and ozone. *Energy Policy* **35**, 5370–5383 (2007).
28. Paterson, R. R. M. & Lima, N. How will climate change affect mycotoxins in food? *Food Research International* **43**, 1902–1914 (2010).
29. James, S. & James, C. The food cold-chain and climate change. *Food Research International* **43**, 1944–1956 (2010).
30. Parfitt, J., Barthel, M. & Macnaughton, S. Food waste within food supply chains: quantification and potential for change to 2050. *Philosophical Transactions of the Royal Society B: Biological Sciences* **365**, 3065–3081 (2010).

11
Nitrogen and Biofuels

Biofuels are fuels derived directly or indirectly from biological material and, in their simplest forms, have been humankind's longest-running method of energy management. Since our ancestors first harnessed the heating and lighting power of fire, burning biomass has been a cornerstone of human civilisation. As the industrial revolution spread around the world in the 18th and 19th centuries, fossil fuel use took over from bioenergy in many areas. Fossil fuels such as coal, gas and oil packed much more of an energy punch than the same mass or volume of biomass. It was fossil energy, not biomass energy, that was required to power the global expansion of industry, electricity generation and motorised transport.

So far, the boom in fossil fuel burning in much of the Western world over the last two centuries has not been repeated in all countries. Across the developing world bioenergy in the form of wood, charcoal or organic waste remains the primary source of heat for millions of people[1]. Increasing access to and use of fossil fuels was long seen as a key to economic development and improved quality of life, with the stunning speed and scale of development in the oil-rich Gulf states bearing testament to the benefits of exploiting fossil fuels and the sidelining of bioenergy.

Only in the latter half of the 20th century did the global environmental cost of burgeoning fossil fuel burning begin to emerge[2]. The rapid rise in atmospheric carbon dioxide concentrations[3] and the threat of global warming has removed the 'wonder fuels' tag from coal, oil and gas, and has brought attention back to bioenergy. Whether it is replacing current fossil fuel use in some countries[4], or helping to avoid a future fossil fuel-intensive development pathway in others, these biomass-derived fuels are becoming ever more central to policies aimed at addressing the challenges of energy security and climate change[5,6].

Biofuels and bioenergy

The terms 'biofuel' and 'bioenergy' are often used interchangeably, describing the use of biological material (biomass) to generate fuel and energy respectively. 'Biofuel' is most commonly discussed in relation to transport and the substitution of gasoline and diesel, while 'bioenergy' is used as a more all-encompassing term to describe the energy produced from biofuels, including the use of wood and charcoal for heating and electricity generation[7].

In its most traditional form, bioenergy production is simply the burning of wood for heating and cooking. Across the world various other sources of such biomass energy are used, including charcoal and agricultural residues such as straw, rice husks and dried manure. Whatever the biomass fuel used (called the 'feedstock'), the most direct method of unlocking the energy it contains is to burn it.

Biomass can also provide a source of energy indirectly through processing of biological material to generate a more concentrated form of fuel. The most common fuel of this type is ethanol derived from the fermentation of agricultural crops such as maize, wheat and sugar cane. In this process the sugars in the crops are converted to alcohol which is then used as a substitute for liquid fossil fuels like gasoline. Oil-producing crops like oilseed rape, soybeans and sunflowers are also used widely to produce a biofuel called 'biodiesel', where the oils are converted by a process called 'transesterification' to a form similar to the fossil diesel burned by many vehicles[8]. The main feature these bioethanol and biodiesel fuels have in common is that they are all derived from agricultural food crops, and these fuels are collectively known as 'first-generation biofuels'[9].

A major criticism of these first-generation biofuels has been that, as they rely on using agricultural crops, they compete with food production and so can push up food prices and exacerbate food security problems. One solution to this problem is to instead use feedstocks that come from agricultural and forest wastes or those that are derived from non-food crops for biofuel production. These 'second-generation biofuels'[10] include biodiesel produced from inedible oil-producing plants like *Jatropha* and bioethanol obtained by conversion of straw, woody clippings and grasses using enzymes to break down the large amounts of lignin and cellulose in such material into simple sugars[11]. The problem with the latter approach is that the rate at which these 'lignocellulosic' feedstocks can be converted to sugars suitable for alcohol production tends to be slow, and producing the very large amounts of bioethanol required to substitute fossil fuel use globally by this method remains out of reach. A breakthrough is required that allows faster and more cost-effective conversion of these hard-to-break-down lignocellulosic feedstocks. Current research in this area is focussed on developing more energy-efficient processing methods and on identifying microbes that could be especially effective at breaking down plant cellulose into useful products. A fungi called *Gliocladium roseum* that was isolated from the forests of Patagonia appears to be able to do just this, converting woody material to products similar to those found in diesel fuel[12]. Similarly, a common bacterium called *Escherichia coli* has been genetically modified to enable it to produce alcohols or biodiesel[13,14], prompting further optimism that engineered microorganisms can be used to improve the net climate benefits of biofuels[15].

Fossil fuel versus bioenergy

Each year the burning of fossil fuels leads to the emission of over eight billion tonnes of carbon into the atmosphere in the form of carbon dioxide[16]. It takes many millions of years for fossil carbon such as coal and oil to be formed, with coal deposits mainly derived from ancient forests and oil reserves resulting from the prehistoric algal blooms that formed ocean sediments.

Given the vast time span required to reincorporate atmospheric carbon dioxide into coal and oil, any carbon emitted when fossil fuels are burned is effectively a net loss from the world's fossil carbon stock. The 40 per cent increase in atmospheric concentrations of carbon dioxide since the industrial revolution has largely been a result of such fossil fuel burning[16], with growing energy demand and use driving emissions higher and higher each year.

Bioenergy is also a source of carbon dioxide, with the wood, straw or other biomass energy source emitting carbon dioxide when it is burned. Here though, the emitted carbon dioxide is regarded as being in balance with that taken up by plants for new biomass growth on the land and in the ocean. As such, bioenergy is commonly regarded as 'carbon neutral', with carbon emissions from combustion assumed to be balanced by carbon uptake via photosynthesis.

Climate change and biofuels

The climate change mitigation benefits of biofuels seem obvious, given that they are regarded as carbon neutral and can substitute for carbon-intensive fuels such as gasoline. One major issue with this assumption is the impact of expanding forest clearance and land cultivation for biofuel crops[17]. Converting forests and unmanaged land to agriculture can result in large net losses of carbon to the atmosphere as trees are removed and the soil is ploughed up. If this conversion is speeded up by a growing demand for biofuel crops, the net effect of biofuels on the climate may be much less beneficial than the simple substitution of fossil fuels with ethanol or biodiesel would suggest. This problem is especially important for first-generation biofuels where the crops are planted specifically for fuel production. This requires either new land to be cultivated or for croplands previously used for food production to be pushed into new areas. Either way, it can incur a substantial additional 'carbon penalty' for the biofuels through the drastic change in land use[18].

Properly accounting for the land-use change emissions of biofuel production is vital when deciding what climate change mitigation benefits, if any, a particular biofuel has. Potentially even more important than this element though is the role of nitrogen in biofuel production: the energy use associated with nitrogen fertiliser production and just how much of the powerful greenhouse gas nitrous oxide is emitted when these fertilisers are applied[19].

Nitrous oxide and biofuels

Just as crops grown for food often require large additions of nitrogen fertiliser, so too do crops grown for biofuel production. Once added to the soil this nitrogen can drive faster plant growth, but it can also lead to larger emissions of nitrous oxide through the processes of nitrification and denitrification. Precisely how much nitrous oxide is emitted from a biofuel crop depends on the amount of nitrogen fertilisation, soil conditions and the type of crop itself. Where nitrous oxide emissions are large, they have the potential to reduce or completely offset the climate change mitigation benefits achieved by substituting fossil fuel burning with biofuels[20].

Again, it is the cultivation of first-generation biofuel crops like maize and sugar cane that runs the greatest risk of incurring a heavy nitrous oxide penalty. The danger is that national and global policies that focus too much on reducing carbon emissions from energy and industry will incentivise vast increases in bioenergy production even where the nitrous oxide emissions make them a poor tool for climate change mitigation[21]. Global production of wheat, coarse grains and vegetable oils for biofuel use, for example, is projected to rise from around 160 million tonnes in 2010 to over 200 million tonnes by 2020[22], and big questions therefore remain on how much expansion in cultivated land area and what increase in global nitrous oxide emissions this biofuel boom will lead to[23] (Box 11.1).

Box 11.1

Corn ethanol and climate change

In 2006 the US produced almost half of the world's total biofuels, largely using a feedstock of corn (also called maize or *Zea mays*) to produce ethanol. It produced 18 billion litres of ethanol that year, with this figure set to rise to 57 billion litres by 2015[24]. By 2007 ethanol was being blended into around half of all US gasoline to produce a mix of 10 per cent ethanol and 90 per cent gasoline (called E10), with the objective of raising this blend ratio to produce a mix of 85 per cent ethanol and 15 per cent gasoline (called E85) in future years[25].

President George W. Bush then announced a major rise in US biofuel production targets with the Energy Independence and Security Act (EISA)[26]. The act aimed to reduce the US dependence on oil by expanding the production of renewable fuels and making the country 'cleaner for future generations'. It planned to do so by increasing the amount of biofuel production to 136 billion litres a year by 2022, with more than half of this coming from advanced second-generation biofuels such as lignocellulosic ethanol.

The debate over whether first-generation biofuels, and especially corn ethanol, do significantly reduce greenhouse gas emissions relative to the fossil fuels they are replacing has become ever-more fierce as their production has increased. One review of biofuel studies found that of the 13 feedstocks analysed, 'all biofuels, except corn ethanol, show strong potential to reduce pollution and cut CO_2 outputs'[27]. Another study reported that corn ethanol could cut greenhouse gas emissions by ~40 per cent compared to gasoline[28], but that this fell far short of other non-food crop feedstocks such as reed canary grass, switch grass and poplar. When the global land-use change emissions of corn ethanol production were also examined, the climate change mitigation benefits appeared to be even less attractive; one study estimated that, compared to gasoline, corn ethanol could actually increase greenhouse gas emissions by over 90 per cent[17].

A major uncertainty in all studies of first-generation biofuel use for climate change mitigation is the degree to which nitrous oxide emissions are increased by growing the feedstock. The standard approach is to estimate nitrous oxide emissions based on the Intergovernmental Panel on Climate Change (IPCC) default emission factors[29]. These assume that for every tonne of nitrogen fertiliser used to grow the biofuel crop, just over one per cent is emitted as nitrous oxide. Then, in a landmark study in 2008[21], Nobel Laureate Paul Crutzen and colleagues suggested that the proportion of nitrogen inputs emitted as nitrous oxide (called the 'emission factor') may actually be much higher, at between three per cent and five per cent. If correct, this would mean that the nitrous oxide penalty inherent in the production of many first-generation biofuels would offset a much larger part of any carbon reduction benefits.

Box 11.2

Sugar cane and nitrous oxide

Like corn, an increasing amount of sugar cane cultivation is being devoted to the production of ethanol as a biofuel. The bulk of the world's sugar cane crop for ethanol production is in Brazil, with an output of over 20 billion litres each year. Brazilian sugar cane has a distinct climate change mitigation advantage over American corn ethanol in that its cultivation is predominantly on existing agricultural land and does not lead to additional land-use change. As such, carbon dioxide emissions derived from land-use change tend to be very low. However, sugar cane is a nitrogen-intensive crop, and the relatively high temperatures and soil water contents in the tropics can enhance denitrification rates, boosting emissions of nitrous oxide. Though reported emissions from Brazilian sugar cane production appear low, recent studies examining sugar cane production in Australia have reported very high emission factors (the proportion of nitrogen applied as fertiliser that is emitted as nitrous oxide). In these areas an emission factor of three to five per cent appears more common to sugar cane cultivation – a very similar figure to that suggested by Paul Cruzen and his colleagues[21]. At present, the global coverage of field measurements of nitrous oxide fluxes from biofuel cultivation is still not sufficient to allow reliable estimation of fluxes for all regions, land-management practices and crops. It may well be that the IPCC's default emission factor of one per cent for emissions arising directly from nitrogen addition is broadly representative for sugar cane produced in Brazil. But if an emission factor of three to five per cent is more representative for other sugar cane systems around the world, the nitrous oxide 'penalty' of global sugar cane production for biofuels may be much higher than was thought previously[30].

Using this new Crutzen emission factor of three to five per cent instead of the standard IPCC emission factor of around one per cent means that by 2020 the projected corn ethanol production in the US (Box 11.1) goes from achieving a 30 per cent cut in net greenhouse gas emissions, compared to gasoline, to as little as a four per cent cut. In this case the higher nitrous oxide emissions effectively wipe out all of the climate change mitigation benefits of corn ethanol[25]. On this evidence, if the US is to produce biofuels in a more environmentally sustainable manner, greater use of non-food crop feedstocks that require lower nitrogen inputs would be needed. Indeed, wherever in the world climate change mitigation is the primary objective of biofuel cultivation, the fullest possible picture in terms of net climate forcing is needed before policy decisions are made (Box 11.2).

Second- and third-generation biofuels

As well as avoiding increased pressures on land use and food prices, second-generation biofuels have the potential for substantial net climate change mitigation as they can avoid some or all of the deliberate nitrogen inputs[19] and resulting nitrous oxide emissions common to first-generation biofuels. Much attention and research is now focused on the growth of non-food 'super biofuel' feedstocks that provide products that are easy to grow and convert into usable fuel. One such feedstock is the fast-growing shrub *Jatropha*, which can give very high oil yields from marginal land that could not support food crop production. Because *Jatropha* has good drought and pest resistance it may be especially useful in low-rainfall areas where agriculture is already threatened by the impacts of climate change[31]. As it produces berries with an oil content of over 30 per cent, it could provide sizeable increases in biodiesel production and be a valuable source of income for subsistence farmers in the developing world. Like *Jatropha*, the fast-growing grass *Miscanthus* has also received a great deal of interest because of its potential as a large-scale biofuel source[32]. *Miscanthus* can grow by more than three metres in just one year, with the large amounts of biomass produced then being used either for ethanol production or for direct burning for heat and electricity generation. Again, it requires little or no nitrogen fertilisers and so avoids the large nitrous oxide emissions that can occur through producing corn ethanol and other first-generation biofuels. Low or zero reactive nitrogen inputs also help to avoid the problems of nitrate leaching, water pollution and downstream eutrophication common to many intensive agricultural systems around the world (see Chapter 7).

Algal biofuels – The third generation

Some non-food biofuel feedstock supplies could even dispense with the need for land cultivation altogether[33]. In the production of what is sometimes referred to as 'third-generation' biofuels[34], the large-scale growing of algal biomass can benefit from very fast growth rates and manipulation of the types of algae grown to deliver the best possible yields of oil for biodiesel production[35]. Though still not produced on a commercial scale, algal biofuels of this type could be produced using nutrient-rich wastewater, and thereby avoid the land- and water-use costs of other biofuel feedstocks[36]. In theory, the combination of a site with plentiful sunlight and good access to wastewater supplies could generate a 100-fold increase in biofuel from algae than would be possible with more traditional biofuel feedstocks[37]. The algae can be grown in vertical incubators to reduce the land footprint, with their extremely rapid growth rates allowing huge amounts of algal biomass to be produced very quickly. The large-scale production of algal biomass under these controlled conditions, and its subsequent conversion to biodiesel or ethanol, holds the promise of avoiding land-use change and food-price penalties while delivering significant substitution of fossil fuels globally. Importantly, such controlled production of algal biomass could also avoid the large nitrous oxide penalties inherent in the production of some first-generation biofuels. By growing algae under optimal conditions of light, temperature and nutrient supply, their nitrogen use efficiency can be maximised and the loss of reactive nitrogen via denitrification (as nitrous oxide and dinitrogen gas) minimised.

The main barrier to such a global spread of algal biofuel production remains the high cost of the incubation equipment, the relatively low biomass concentrations in the cultures (at higher densities the algae start to shade each other) and the subsequent processing into usable biofuels[38]. If a combination of more efficient incubators and genetic manipulation of microbes can reduce these costs, then third-generation biofuels may yet be the long-term replacement fuel for the world's cars, trucks and even planes.

References

1. Rivard, B. & Reay, D. S. Future scenarios of Malawi's energy mix and implications for forest resources. *Carbon Management* 3, 369–381, doi:10.4155/cmt.12.35 (2012).
2. Solomon, S. *Climate change 2007 – the physical science basis: Working Group I contribution to the fourth assessment report of the IPCC.* Vol. 4 (Cambridge University Press, 2007).

3. Keeling, C., Whorf, T., Wahlen, M. & Plicht, J. v. d. Interannual extremes in the rate of rise of atmospheric carbon dioxide since 1980. *Nature* **375**, 666–670 (1995).
4. Berndes, G. & Hansson, J. Bioenergy expansion in the EU: cost-effective climate change mitigation, employment creation and reduced dependency on imported fuels. *Energy Policy* **35**, 5965–5979 (2007).
5. Berndes, G., Hoogwijk, M. & van den Broek, R. The contribution of biomass in the future global energy supply: a review of 17 studies. *Biomass and Bioenergy* **25**, 1–28 (2003).
6. Smith, P. et al. Greenhouse gas mitigation in agriculture. *Philosophical Transactions of the Royal Society B: Biological Sciences* **363**, 789–813 (2008).
7. Group, U. N. E. P. B. W. & Management, U. N. E. P. I. P. f. S. R. *Towards sustainable production and use of resources: assessing biofuels.* (UNEP/Earthprint, 2009).
8. Hill, J., Nelson, E., Tilman, D., Polasky, S. & Tiffany, D. Environmental, economic, and energetic costs and benefits of biodiesel and ethanol biofuels. *Proceedings of the National Academy of Sciences of the United States of America* **103**, 11206–11210 (2006).
9. Naik, S., Goud, V. V., Rout, P. K. & Dalai, A. K. Production of first and second generation biofuels: a comprehensive review. *Renewable and Sustainable Energy Reviews* **14**, 578–597 (2010).
10. Sims, R. E., Mabee, W., Saddler, J. N. & Taylor, M. An overview of second generation biofuel technologies. *Bioresource Technology* **101**, 1570–1580 (2010).
11. Somerville, C., Youngs, H., Taylor, C., Davis, S. C. & Long, S. P. Feedstocks for lignocellulosic biofuels. *Science (Washington)* **329**, 790–792 (2010).
12. Strobel, G. A. et al. The production of myco-diesel hydrocarbons and their derivatives by the endophytic fungus Gliocladium roseum (NRRL 50072). *Microbiology* **154**, 3319–3328 (2008).
13. Steen, E. J. et al. Microbial production of fatty-acid-derived fuels and chemicals from plant biomass. *Nature* **463**, 559–562 (2010).
14. Bokinsky, G. et al. Synthesis of three advanced biofuels from ionic liquid-pretreated switchgrass using engineered Escherichia coli. *Proceedings of the National Academy of Sciences of the United States of America* **108**, 19949–19954 (2011).
15. Singh, B. K., Bardgett, R. D., Smith, P. & Reay, D. S. Microorganisms and climate change: terrestrial feedbacks and mitigation options. *Nature Reviews Microbiology* **8**, 779–790, doi:10.1038/nrmicro2439 (2010).
16. Stocker, T. *Climate change 2013: the physical science basis: Working Group I contribution to the Fifth assessment report of the Intergovernmental Panel on Climate Change.* (Cambridge University Press, 2014).
17. Searchinger, T. et al. Use of US croplands for biofuels increases greenhouse gases through emissions from land-use change. *Science* **319**, 1238–1240 (2008).
18. Fargione, J., Hill, J., Tilman, D., Polasky, S. & Hawthorne, P. Land clearing and the biofuel carbon debt. *Science* **319**, 1235–1238 (2008).
19. Erisman, J. W., van Grinsven, H., Leip, A., Mosier, A. & Bleeker, A. Nitrogen and biofuels; an overview of the current state of knowledge. *Nutrient Cycling in Agroecosystems* **86**, 211–223 (2010).
20. Smeets, E. M. et al. Contribution of N2O to the greenhouse gas balance of first-generation biofuels. *Global Change Biology* **15**, 1–23 (2009).

21. Crutzen, P. J., Mosier, A. R., Smith, K. A. & Winiwarter, W. N2O release from agro-biofuel production negates global warming reduction by replacing fossil fuels. *Atmospheric Chemistry and Physics* **8**, 389–395 (2008).
22. Alexandratos, N. & Bruinsma, J. World agriculture towards 2030/2050: the 2012 revision. (ESA Working Paper Rome, FAO, 2012).
23. Galloway, J. N. et al. Transformation of the nitrogen cycle: recent trends, questions, and potential solutions. *Science* **320**, 889–892, doi:10.1126/science.1136674 (2008).
24. Leonard, B. *USDA Agricultural Projections To 2017*. (DIANE Publishing, 2011).
25. Galt, H. & Reay, D. S. Corn ethanol and associated greenhouse gas emissions in the USA: importance of the nitrous oxide emission factor. *Carbon Management* **2**, 13–22, doi:10.4155/cmt.10.38 (2011).
26. Sissine, F. Energy Independence and Security Act of 2007: a summary of major provisions. *CRS Report for Congress RL34294*, 22 pp. (2007).
27. Groom, M. J., Gray, E. M. & Townsend, P. A. Biofuels and biodiversity: principles for creating better policies for biofuel production. *Conservation Biology* **22**, 602–609 (2008).
28. Adler, P. R., Del Grosso, S. J. & Parton, W. J. Life-cycle assessment of net greenhouse-gas flux for bioenergy cropping systems. *Ecological Applications* **17**, 675–691 (2007).
29. Eggleston, S., Buendia, L., Miwa, K., Ngara, T. & Tanabe, K. IPCC guidelines for national greenhouse gas inventories. *Institute for Global Environmental Strategies, Hayama, Japan* (2006).
30. Reay, D. S. Not so sweet after all? *Nature Climate Change* **1**, 174 (2011).
31. Makkar, H. P. & Becker, K. Jatropha curcas, a promising crop for the generation of biodiesel and value-added coproducts. *European Journal of Lipid Science and Technology* **111**, 773–787 (2009).
32. Davis, S. C. et al. Comparative biogeochemical cycles of bioenergy crops reveal nitrogen-fixation and low greenhouse gas emissions in a Miscanthus × giganteus agro-ecosystem. *Ecosystems* **13**, 144–156 (2010).
33. Mata, T. M., Martins, A. A. & Caetano, N. S. Microalgae for biodiesel production and other applications: a review. *Renewable and Sustainable Energy Reviews* **14**, 217–232 (2010).
34. Brennan, L. & Owende, P. Biofuels from microalgae – a review of technologies for production, processing, and extractions of biofuels and co-products. *Renewable and Sustainable Energy Reviews* **14**, 557–577 (2010).
35. Schenk, P. M. et al. Second generation biofuels: high-efficiency microalgae for biodiesel production. *Bioenergy Research* **1**, 20–43 (2008).
36. Pittman, J. K., Dean, A. P. & Osundeko, O. The potential of sustainable algal biofuel production using wastewater resources. *Bioresource Technology* **102**, 17–25 (2011).
37. Greenwell, H., Laurens, L., Shields, R., Lovitt, R. & Flynn, K. Placing microalgae on the biofuels priority list: a review of the technological challenges. *Journal of the Royal Society Interface*, rsif20090322 (2009).
38. Carriquiry, M. A., Du, X. & Timilsina, G. R. Second generation biofuels: economics and policies. *Energy Policy* **39**, 4222–4234 (2011).

12
Nitrogen and Geoengineering

Proposals for the deliberate manipulation of Earth's climate to tackle human-induced global warming – popularly named 'geoengineering' – have grown in number as fast as the frustration over continually rising greenhouse gas emissions has intensified. Where the line between conventional climate change 'mitigation' and 'geoengineering' is drawn varies from study to study, but here we consider it to be any single strategy that could deliver an annual change in global climate forcing equivalent to removing more than one billion tonnes of carbon (1 Pg C) from the atmosphere. To put this one billion tonnes of carbon in context, it is equivalent to over 3.6 billion tonnes of carbon dioxide or about one-tenth of global carbon dioxide emissions from fossil fuel burning each year.

All the proposed strategies can be divided broadly into two types. The first is called 'carbon dioxide reduction' (CDR) and encompasses large-scale reductions in the carbon dioxide concentration in the atmosphere that serve to reduce or completely reverse the impacts of anthropogenic greenhouse gas emissions. The second is called 'solar radiation management' (SRM) and involves the amount of energy received or reflected by the Earth being altered[1]. Nitrogen could play an important direct or indirect role in many of these CDR and SRM proposals, its complex cycling through atmosphere, water, soil and biomass making for numerous potential geoengineering synergies and antagonisms (Figure 12.1).

Carbon dioxide reduction

Many CDR strategies for geoengineering appear more attractive than SRM options, as they can address the root cause of human-induced warming – elevated greenhouse gas concentrations in the atmosphere – as well as the warming itself. Proposed strategies range from very large-scale

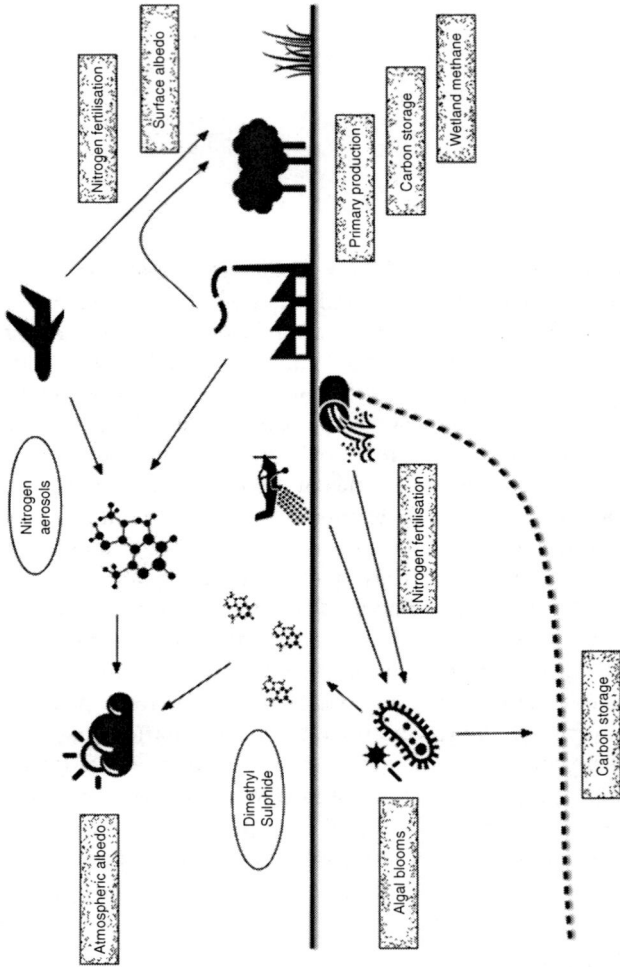

Figure 12.1   Nitrogen and geoengineering

The solid line represents the land and ocean surface and the dotted line the ocean floor. The boxes show the mechanisms by which nitrogen might play a role in SRM or in CDR. Potential mechanisms for SRM involve changing atmospheric reflectance (albedo) through nitrogen aerosols and dimethyl sulphide emissions from algal blooms, and changing surface albedo through changing vegetation growth. Possible ones for CDR involve increasing primary production on land or in the oceans through nitrogen fertilisation.

Source: Dave Reay

tree planting to 'artificial trees' that absorb carbon dioxide from the air for subsequent storage underground, to manipulating ocean circulation to sequester more carbon in the deep ocean[2]. As primary production over much of the Earth's land and sea is nitrogen-limited, deliberate nitrogen fertilisation also has the potential to increase carbon dioxide uptake from the atmosphere. If achieved at a large enough scale, nitrogen fertilisation could therefore address some or all of the enhanced warming and acidification problems caused by rising carbon dioxide concentrations.

The main proposals aimed at achieving geoengineering-type reductions in greenhouse gas concentrations in the atmosphere revolve around promotion of algal growth and carbon sequestration in the oceans. For the 'high-nutrient low-chlorophyll' (HNLC) oceanic regions, such as the equatorial Pacific and the Southern Ocean, increasing the availability of iron (a key micronutrient) has successfully been used to promote large algal blooms[3,4]. For other areas of the world's oceans, where nitrogen can be a key limiting factor (Box 12.1), deliberately increasing nitrogen supplies to algae in surface waters could bring about the same bloom effect.

Changing the amount of carbon taken up by algae in the world's oceans therefore depends upon manipulating their access to the key nutrients they require for growth. For most algae the required ratio of carbon to nitrogen is 106:16. As nitrogen supply is limited in much of the open ocean, every atom of added nitrogen could in theory help the algae to sequester around six atoms of carbon. Realising such a big response for the purposes of geoengineering depends on the extent of limitation by other nutrients (called 'co-limitation'), how easily the extra nitrogen can be applied in the right place at the right time and just how long any extra carbon taken up by this route stays in the oceans and out of the atmosphere.

Oceanic nitrogen fertilisation

In the many areas of the ocean where nitrogen limits algal growth, the deliberate addition of extra nitrogen to induce big algal blooms, and so sequester extra carbon from the atmosphere, may seem a viable geoengineering option. However, there are numerous major barriers to its implementation and big questions as to the overall effectiveness of this strategy[4]. The first consideration is how to deliver the vast amounts of reactive nitrogen required to these oft-remote ocean areas. To make a meaningful dent in atmospheric carbon dioxide concentrations would require an increase in net ocean carbon uptake of several billion tonnes per year – current carbon emissions from fossil fuel burning are around eight billion tonnes a year. To induce enhanced algal growth equivalent

Box 12.1

Ocean nitrogen boom and bust

Marine algae are only able to capture enough sunlight for photosynthesis in the top 100–200 metres of the ocean (the photic zone), and it is in this narrow zone that nitrogen supplies can be rapidly exhausted, with further algal growth and carbon sequestration becoming nitrogen-limited. Such exhaustion of supplies is a common part of the seasonal cycle of algal boom and bust in temperate oceans, and is a perennial cap on primary production across much of the open ocean in the tropics.

In temperate oceans the foundations of the boom phase are built in winter. A combination of low water temperatures and raging storms allow better mixing of deep nutrient-rich waters with those at the surface. With little daylight, and therefore very slow algal growth, nitrogen stocks in the surface waters are replenished.

Then, with the onset of spring, water temperatures and light availability begin to increase. The surface waters become more stable and the algae are able to take advantage of the plentiful nutrient supplies to grow fast and produce big peaks in primary production and carbon uptake. Unfortunately for the algae, this boom time is short-lived. The stratification of surface waters that occurs as they warm up in late spring and summer means that relatively little nitrogen can be supplied from the deeper, nutrient-rich waters. As nitrogen supplies dwindle, algal numbers slump and carbon uptake falls away. Throughout the rest of the summer the oceanic algae must then rely on whatever nitrogen is deposited from the atmosphere, fixed by marine organisms like the cyanobacteria or recycled via decomposition and excretion from grazers.

Only when the first storms of autumn roll in and the stratification of the ocean layers is broken down can the algae once again get access to large supplies of nitrogen. If water temperatures and amounts of sunlight are still high enough, then a second big boom in primary production can occur before the cold and dark days of winter again strangle algal growth. Climate change is expected to alter the timings of this oceanic boom and bust, with higher temperatures allowing an earlier start to the spring bloom in some areas, but with changes to wind and storm patterns affecting where and when the nutrient-rich deep ocean waters become mixed with those at the surface.

(Continued)

Box 12.1   (Continued)

In tropical seas, it is strong stratification – the poor mixing of the colder, deep waters with the much warmer, sunlit surface waters – that keeps a tight reign on algal growth and makes for the low levels of primary production seen over very large areas of the tropical ocean. In the few places where nitrogen-rich deep waters are able to upwell to the surface in the tropics, the surge in algal growth and the boom along the whole marine food chain can be stunning. Just such a tropical ocean upwelling zone is off the west coast of Peru, where the flood of nutrients into the warm, well-lit waters produce a sustained algal bloom that supports one of the most productive fisheries in the world. It was the huge abundance of fish in this area and the many centuries of harvesting by seabirds that led to the build-up of the 'guano mountains' on shore that were so viciously fought over in the 19th century for the precious nitrogen they contained (see Chapter 2). Today the guano mountains are all gone, but many thousands of people still rely on the productivity of the Peru upwelling zone for their livelihoods. When the upwelling fails, as happens every seven years or so, the consequences for the marine food chain and the humans who rely on it can be severe. The wider cause of this failure is called El Niño and it involves a weakening or reversal of the strong easterly winds that normally blow across the Pacific Ocean. Off the coast of Peru these easterly winds are constantly pushing the surface waters away from the land, thereby allowing the nitrogen-rich deep water to well-up and replace them. As the winds falter, the warm surface waters that had been pushed away to the west can creep back east towards South America. They effectively form a cap of stratified water that shuts off the deep water upwelling, causing algal growth in the region to crash – and with it the hugely productive fisheries that it supports. El Niño, or El Niño Southern Oscillation (ENSO) to give it its full name, may itself be affected by climate change, with changes in wind patterns and ocean currents altering the frequency and intensity of these events[5].

to just one billion tonnes of carbon uptake would require at least 167 million tonnes of reactive nitrogen to be delivered to the ocean at exactly the right time and place every year. The costs of such year-round application via aircraft would be prohibitive, while delivery via ships would limit the amount of ocean area that could be covered and would also incur astronomical costs. Delivery to the sea bed by pipelines or barrels that provide a gradual seep of reactive nitrogen might help avoid the need for continual reapplication to the same areas each year, but such systems would be ineffective in the open ocean areas where nitrogen limitation is most critical[6]. Even if a more efficient delivery method were developed, the reactive nitrogen is itself a precious resource in terms of agricultural production, and its use for ocean fertilisation would incur an enormous additional cost. With the current cost of nitrogen fertiliser at ~$1,000 per tonne, applying 167 million tonnes of fertiliser to the oceans for geoengineering would cost more than $160 billion per year even before the application costs were added in.

The key assumption that every bit of the added nitrogen would power additional algal growth is also flawed. Though nitrogen is limited in large areas of the ocean, phosphorous is also in short supply. This means that in up to 70 per cent of the ocean, relieving nitrogen limitation will only result in phosphorous limitation of algal growth taking over[7]. In fact, in most nitrogen-limited areas of the world's oceans, natural nitrogen fixation from the atmosphere is a major source of nitrogen for algal growth[8,9]. Deliberate additions would likely decrease this natural fixation and the net benefit in terms of algal growth would therefore be small.

One alternative suggestion to adding new nitrogen to the oceans has been the deliberate enhancement of upwelling of nutrient-rich deep waters to the surface. Certainly, if responses such as that seen off the coast of Peru could be replicated, the creation of large algal blooms might be expected. To achieve this 'artificial upwelling' a system of long vertical pipes has been suggested, whereby nutrient-rich waters are pumped to the ocean surface from several hundred metres below[10]. The feasibility of such an approach over the area required for geoengineering is doubtful. More importantly, the costs and net effect on carbon sequestration are uncertain[11]. Deliberately bringing these deeper waters to the surface may well relieve nutrient limitation and enhance algal growth, but this deep water may also bring with it large amounts of dissolved carbon dioxide that is then released into the atmosphere from the ocean surface. Using very optimistic assumptions of effectiveness and scale, a deliberate enhancement of ocean upwelling by one million cubic metres per second – a mammoth engineering challenge – has been predicted to induce just 20 million tonnes of additional carbon sequestration each year[12].

It is such poor effectiveness in terms of net carbon sequestration that ultimately undermines all of the ocean fertilisation strategies for geoengineering touted to date. Though these schemes may achieve visually impressive blooms and enhance removal of dissolved carbon dioxide from the surface ocean in the short term, it is the longevity of this removal that is crucial when deciding whether such approaches are worthwhile in mitigating climate change. Unfortunately, the vast majority of such phytoplankton-bound carbon is rapidly cycled back into the dissolved carbon dioxide pool via respiration and microbial decomposition, with much then being re-emitted to the atmosphere. Indeed, in most oceanic areas only a small proportion of the carbon (~1 per cent) initially incorporated into such algal blooms ends up sinking down to deep water sediments where it can more reliably be expected to remain out of the atmosphere for a length of time relevant to reduced climate forcing (centuries to millennia)[1]. To put this poor effectiveness into perspective, if we assume just one per cent of the extra carbon taken up by a nitrogen-induced algal bloom remains sequestered in the ocean in the long term, then to achieve a growth of one billion tonnes per year in the ocean carbon sink we would need a global nitrogen addition of over 16 billion tonnes per year at a cost of over $10 trillion.

Ocean fertilisation side effects

The limited effectiveness and high costs of sequestering carbon through deliberate nitrogen fertilisation of the oceans represent a clear barrier to this strategy ever being employed on a large scale. Less obvious are the uncertainties surrounding the side effects of such enhancement of ocean algal blooms. Oceanic areas already receiving large inputs of nitrogen and other nutrients from the land, such as the Gulf of Mexico, demonstrate how large algal blooms can have devastating impacts on biodiversity and fish stocks through the creation of 'dead zones' (see Chapter 8). Deliberate enhancement of algal blooms may therefore increase the number, size and intensity of these low-oxygen areas. As well as potential damage to fisheries and marine biodiversity, the low-oxygen conditions may increase production and emission of the powerful greenhouse gases methane and nitrous oxide, thereby offsetting the climate-forcing benefits of any enhanced carbon dioxide uptake[13]. The types of algae that make up the artificial blooms can also cause problems, as some produce toxins that can then enter the marine food chain and pose a serious risk to human health[14].

The one major beneficial side effect of artificial ocean algal blooms may be via the indirect effect that marine algae can have on our climate. As well as sequestering carbon, algal blooms emit a gas called dimethyl sulphide (DMS). Large algal blooms can emit very large amounts of DMS; globally, marine algae are the biggest biological source of sulphur to our atmosphere[15]. Once in the atmosphere this gas is converted to compounds such as sulphur dioxide that can go on to increase cloud formation and the overall albedo (reflectance) of the atmosphere. As such, if DMS emissions were greatly increased by enhancing marine algal blooms, this could help reduce solar energy input and thereby limit global warming. Though large uncertainties remain as to how important this 'DMS climate effect' is, it has been suggested as a major feedback mechanism for climate change in the 21st century. If climate change results in enhanced algal growth and a consequent rise in DMS emissions from the ocean, then this may reduce the warming – a negative feedback. However, if climate change reduces marine algal growth and DMS emissions, then global warming may be intensified – a positive feedback. Many ocean climate models indicate that warming will increase stratification in oceanic areas and would therefore reduce algal growth. Perhaps more importantly, the increasing acidity of the oceans caused by rising atmospheric carbon dioxide concentrations is projected to cause a marked drop in DMS emissions from marine algae. By 2100 this ocean acidification effect on algal sulphur emissions may amplify global warming by almost 0.5°C[16].

Terrestrial nitrogen fertilisation

Through increased fossil fuel burning and the expansion of agriculture, humankind has already engineered a huge increase in nitrogen inputs to natural terrestrial ecosystems via atmospheric deposition[17]. Deliberate nitrogen fertilisation has also been used for commercial forests in many countries, with the aim of increasing tree growth and timber production. In recent years this practice has become less common as the high price of nitrogen and costs of application over large areas have made it uneconomical. From a geoengineering perspective, the deliberate addition of nitrogen to forests would have to be undertaken on a huge scale to achieve carbon sequestration enhancements that are large enough to make a significant impact (i.e. more than one billion tonnes of extra carbon storage per year). The effectiveness of such a strategy is undermined by the fact that tree growth in many of the world's forests is not only limited by nitrogen supply, but also by factors such as water availability, temperature and light[18]. Aside from the enormous financial costs of nitrogen addition to forests on a global scale, such additions would also increase soil nitrous oxide emissions, thereby offsetting some or all of the carbon sequestration benefits (see Chapter 6). Other potential side effects would include increases in nitrogen leaching to surface and ground waters, changes in forest species composition and increased vulnerability of trees to pest and disease attack. Though deliberate application of nitrogen fertilisers to commercial forests may still occur where the economics of timber production are favourable, its efficacy for global climate change mitigation appears very limited. One means by which local carbon gains may be much higher and nitrogen costs actually turned into benefits is in the strategic use of vegetation and tree planting for the interception of reactive nitrogen losses from agriculture (see Chapter 13).

Wetland nitrogen fertilisation

The geoengineering of other natural ecosystems with nitrogen to manipulate greenhouse gas fluxes has an even weaker evidence base than that for forests, though in theory the widespread addition of reactive nitrogen to the world's wetland areas could reduce methane emissions by inhibiting microbial methane production (methanogenesis)[19–21]. Wetlands are the largest global source of the greenhouse gas methane – more than 100 million tonnes per year[22] – and so an intervention that radically reduces these emissions could have a climate-forcing impact of the magnitude required for geoengineering. Deliberate addition of reactive nitrogen to rice paddies has already been shown to reduce methane emissions in some areas (see Chapter 9). Here, the anoxic conditions of the waterlogged sediments mimic those found in natural wetlands and these, combined with a good supply of simple carbon compounds (acetate and carbon dioxide), allow methane-producing microbes (methanogens) to flourish. By adding reactive nitrogen other microbes (e.g. denitrifiers) are helped to inhibit or outcompete the methanogens, thereby cutting methane production and emissions to the atmosphere[23]. This practice is relatively common in rice agriculture, where the nitrogen additions are primarily used to increase crop production. However, extending this strategy to natural wetlands would require nitrogen additions across enormous areas – there are approximately five trillion square metres of wetlands globally[24] – and would incur astronomical fertiliser and application costs. Fertilising the world's wetland areas with nitrogen is also likely to increase plant growth[25] and with it the overall amount of carbon available for methanogenesis, which would actually boost methane production[26]. Even without these doubts over methane fluxes, the myriad negative effects on biodiversity, water quality and nitrous oxide fluxes that would result make widespread nitrogen fertilisation of wetlands a wholly unsuitable technique for climate engineering.

Solar radiation management

SRM can involve the interception of the Sun's energy before it reaches the Earth's atmosphere, such as by using mirrors or dust placed in orbit around the Sun[27], enhancing the reflectivity (albedo) of the atmosphere[28] or increasing the albedo of the Earth's surface[29]. Nitrogen has a potential role in the latter two SRM strategies due to the radiative forcing effects of airborne nitrogen and the impacts of terrestrial, aquatic and marine nitrogen on primary production (Box 12.2).

Nitrogen aerosols

Current proposals for SRM in the Earth's atmosphere are dominated by use of sulphate aerosols[32]. These aerosols are attractive for geoengineering purposes because their effectiveness in moderating global temperatures has already been shown following large volcanic eruptions. During the eruption of Mount Pinatubo in 1991, for example, around

---

Box 12.2

Nitrogen and surface albedo

Increasing the reflectivity of the Earth's surface has been suggested in several geoengineering strategies, including ideas such as covering the world's deserts with reflective sheeting, painting the roofs of all buildings white and growing higher albedo crops. For the last strategy, the proposal involves altering the crops and grasses grown over very large swathes of agricultural land to types that have more reflective leaves, such as variegated forms[30]. Here, increased nitrogen inputs may be required to maintain yields in the face of reduced plant growth rates. More widely, increasing nitrogen inputs to natural and managed ecosystems globally has the potential to alter plant growth and increase leaf albedo[31]. In reality, the proposed strategies for increasing surface albedo at a globally significant scale all appear of limited effectiveness and may be accompanied by major negative side effects. The use of higher albedo crops and grasses, for instance, is estimated to deliver a maximum cooling of about one watt per square metre (W m^{-2}), but only if applied to all crop and grassland globally[1,12]. The practical barriers to such an alteration appear insurmountable, with the potential side effects on agricultural productivity, price and resilience poorly understood.

20 million tonnes of sulphur dioxide was injected into the stratosphere and spread around the planet. This then formed sulphuric acid aerosols and, because these aerosols are of the ideal size for reflecting sunlight (0.1–1 microns), the amount of sunlight reflected by the atmosphere was significantly increased. The amount of solar energy received by the Earth's surface and lower atmosphere dropped by 10 per cent, and in the year following the eruption global temperatures fell by around 0.5°C. As such, various strategies have been suggested whereby sulphate aerosols are deliberately injected into the atmosphere to counteract the enhanced warming caused by anthropogenic greenhouse gas emissions[1]. Other substances such as soot and engineered particles have also been suggested for this SRM technique, but nitrogen aerosols are less attractive due to their much more limited impact on atmospheric albedo. Neither nitrogen oxide (NOx) gases nor ammonia – the most common airborne forms of reactive nitrogen – are able to reduce solar energy input directly, although they can go on to form aerosols, such as ammonium nitrate, that increase atmospheric and cloud albedo[33]. In geoengineering terms though, the amount of global cooling achieved is minimal. The large amounts of reactive nitrogen unintentionally introduced to our atmosphere each year by fossil fuel burning and agriculture only manage to achieve a reduction in solar energy input of around 0.1 W m^{-2} through their impacts on atmospheric albedo[34]. To offset the enhanced warming caused by anthropogenic greenhouse gas emissions (currently more than 2.7 W m^{-2}), using nitrate aerosols alone could therefore require a deliberate 30-fold increase. Such a massive manipulation of airborne nitrogen is far beyond what is economically feasible. It would also have potentially disastrous consequences for terrestrial, aquatic and marine ecosystems as this super-fertilised atmosphere returned a rain of nitrate to the Earth's surface (see Chapter 5).

Nitrogen geoengineering

More effective, and carrying much less risk than broadcasting millions of tonnes of nitrogen across the world's natural ecosystems, are the many smaller-scale greenhouse gas mitigation strategies that focus on managed environments, especially agriculture (see Chapter 9). Here we can better utilise our understanding of plants, animals and microbes to manipulate nitrogen inputs in a way that maximises the climate-forcing benefits and minimises the financial, ecological and social costs. The cascade effect of reactive nitrogen[35] through ecosystems means that any single strategy that begins with a huge injection of nitrogen will ultimately result in a host of unforeseen and likely unwanted impacts. Successful nitrogen geoengineering then, is more likely to be achieved through a range of well-understood, smaller-scale mitigation strategies rather than a single 'nitrogen bullet' whose trajectory and environmental ricochets can only be guessed at.

References

1. Shepherd, J. *Geoengineering the climate: science, governance and uncertainty.* (Royal Society, 2009).
2. Vaughan, N. E. & Lenton, T. M. A review of climate geoengineering proposals. *Climatic Change* **109**, 745–790 (2011).
3. Boyd, P. W. et al. A mesoscale phytoplankton bloom in the polar Southern Ocean stimulated by iron fertilization. *Nature* **407**, 695–702 (2000).
4. Strong, A., Chisholm, S., Miller, C. & Cullen, J. Ocean fertilization: time to move on. *Nature* **461**, 347–348 (2009).
5. Collins, M. et al. The impact of global warming on the tropical Pacific Ocean and El Niño. *Nature Geoscience* **3**, 391–397 (2010).
6. Wallace, D. et al. *Ocean fertilization: a scientific summary for policy makers.* (IOC/UNESCO, Paris, France, 2010).
7. Tyrrell, T. The relative influences of nitrogen and phosphorus on oceanic primary production. *Nature* **400**, 525–531 (1999).
8. Gruber, N. & Sarmiento, J. L. Global patterns of marine nitrogen fixation and denitrification. *Global Biogeochemical Cycles* **11**, 235–266 (1997).
9. Duce, R. et al. Impacts of atmospheric anthropogenic nitrogen on the open ocean. *Science* **320**, 893–897 (2008).
10. Lovelock, J. E. & Rapley, C. G. Ocean pipes could help the Earth to cure itself. *Nature* **449**, 403 (2007).
11. Oschlies, A., Pahlow, M., Yool, A. & Matear, R. J. Climate engineering by artificial ocean upwelling: channelling the sorcerer's apprentice. *Geophysical Research Letters* **37**, L04701 (2010).
12. Lenton, T. M. & Vaughan, N. E. The radiative forcing potential of different climate geoengineering options. *Atmospheric Chemistry and Physics* **9**, 5539–5561 (2009).
13. Naqvi, S. et al. Increased marine production of N_2O due to intensifying anoxia on the Indian continental shelf. *Nature* **408**, 346–349 (2000).
14. Lewitus, A. J. et al. Harmful algal blooms along the North American west coast region: history, trends, causes, and impacts. *Harmful Algae* **19**, 133–159 (2012).
15. Arnold, H. E., Kerrison, P. & Steinke, M. Interacting effects of ocean acidification and warming on growth and DMS-production in the haptophyte coccolithophore Emiliania huxleyi. *Global Change Biology* **19**, 1007–1016 (2013).
16. Six, K. D. et al. Global warming amplified by reduced sulphur fluxes as a result of ocean acidification. *Nature Climate Change* **3**, 975–978 (2013).
17. Dentener, F. et al. Nitrogen and sulfur deposition on regional and global scales: a multimodel evaluation. *Global Biogeochemical Cycles* **20**, GB4003 (2006).

18. Reay, D. S., Dentener, F., Smith, P., Grace, J. & Feely, R. A. Global nitrogen deposition and carbon sinks. *Nature Geoscience* **1**, 430–437, doi:10.1038/ngeo230 (2008).
19. Granberg, G., Sundh, I., Svensson, B. & Nilsson, M. Effects of temperature, and nitrogen and sulfur deposition, on methane emission from a boreal mire. *Ecology* **82**, 1982–1998 (2001).
20. Balderston, W. & Payne, W. Inhibition of methanogenesis in salt marsh sediments and whole-cell suspensions of methanogenic bacteria by nitrogen oxides. *Applied and Environmental Microbiology* **32**, 264–269 (1976).
21. Lindau, C. Methane emissions from Louisiana rice fields amended with nitrogen fertilizers. *Soil Biology and Biochemistry* **26**, 353–359 (1994).
22. Reay, D. S., Smith, K. A. & Hewitt, C. N. Methane: importance, sources and sinks. In *Greenhouse Gas Sinks*, edited by D. Reay et al., 143–151, doi:10.1079/9781845931896.0143 (2007).
23. Klüber, H. D. & Conrad, R. Inhibitory effects of nitrate, nitrite, NO and N2O on methanogenesis by Methanosarcina barkeri and Methanobacterium bryantii. *FEMS Microbiology Ecology* **25**, 331–339 (1998).
24. Matthews, E. & Fung, I. Methane emission from natural wetlands: global distribution, area, and environmental characteristics of sources. *Global Biogeochemical Cycles* **1**, 61–86 (1987).
25. Verhoeven, J. & Schmitz, M. Control of plant growth by nitrogen and phosphorus in mesotrophic fens. *Biogeochemistry* **12**, 135–148 (1991).
26. Whiting, G. & Chanton, J. Primary production control of methane emission from wetlands. *Nature* **364**, 794–795 (1993).
27. Lunt, D. J., Ridgwell, A., Valdes, P. J. & Seale, A. 'Sunshade World': a fully coupled GCM evaluation of the climatic impacts of geoengineering. *Geophysical Research Letters* **35**, L12710 (2008).
28. Wigley, T. M. A combined mitigation/geoengineering approach to climate stabilization. *Science* **314**, 452–454 (2006).
29. Irvine, P. J., Ridgwell, A. & Lunt, D. J. Climatic effects of surface albedo geoengineering. *Journal of Geophysical Research: Atmospheres (1984–2012)* **116** D24112 (2011).
30. Ridgwell, A., Singarayer, J. S., Hetherington, A. M. & Valdes, P. J. Tackling regional climate change by leaf albedo bio-geoengineering. *Current Biology* **19**, 146–150 (2009).
31. Ollinger, S. et al. Canopy nitrogen, carbon assimilation, and albedo in temperate and boreal forests: functional relations and potential climate feedbacks. *Proceedings of the National Academy of Sciences of the United States of America* **105**, 19336–19341 (2008).
32. Crutzen, P. J. Albedo enhancement by stratospheric sulfur injections: a contribution to resolve a policy dilemma? *Climatic Change* **77**, 211–220 (2006).
33. Pilinis, C., Pandis, S. N. & Seinfeld, J. H. Sensitivity of direct climate forcing by atmospheric aerosols to aerosol size and composition. *Journal of Geophysical Research: Atmospheres (1984–2012)* **100**, 18739–18754 (1995).
34. Solomon, S. *Climate change 2007 – the physical science basis: Working Group I contribution to the fourth assessment report of the IPCC*. Vol. 4 (Cambridge University Press, 2007).
35. Galloway, J. N. et al. The nitrogen cascade. *Bioscience* **53**, 341–356 (2003).

13
Nitrogen and Climate Change Adaptation

Climate change adaptation is the process whereby something or someone's ability to cope with climate change is increased. In the built environment this could involve construction of higher flood defences or larger drains, while in the transport sector it might entail the use of more heat-resistant road surfaces or the design of alternative travel plans for use in the event of landslides.

Climate change is already happening, and its impacts are projected to become increasingly severe as the 21st century progresses[1]. Despite ongoing efforts to reach global agreement addressing greenhouse gas emissions, an increase in global average temperature of more the 2°C relative to the pre-industrial average is likely. As a result, even with aggressive mitigation of climate change a huge amount of adaptation will be required in every sector and every country. It is in the sectors where nitrogen is most is used, such as agriculture, and the environments in which this nitrogen is then processed, that adaptation to climate change holds the greatest risks and rewards.

Adaptation in agriculture

Agricultural production in most areas of the world is threatened by climate change[2]. Though some high-latitude regions may benefit from increased temperatures and longer growing seasons over the next few decades, all regions are expected to see increasingly negative climate change impacts in the second half of the 21st century. Key climate change impacts on agricultural production include an increase in the frequency and severity of droughts, heatwaves, floods and storms, together with indirect threats such as pest, disease and ozone damage[3]. The adaptation response will vary depending on the type of threat, how large it is perceived to be and, often most importantly, the resources available with which to act[4,5].

Agriculture is a major source of reactive nitrogen losses to the air and to aquatic systems (see Chapters 5 and 7). As well as being costly for farmers, these losses can cause a cascade of negative effects ranging from lost biodiversity and increased eutrophication in natural ecosystems to poor water quality and air pollution threats to humans[6–8]. Much effort has been put into reducing these losses and increasing the efficiency of agricultural nitrogen use. Yet big losses, and their consequent problems, still occur, with climate change impacts expected to exacerbate this issue.

Climate resilient crop production and nitrogen

For arable agriculture in much of the developing world, climate change represents a major threat[9]. Where yields are threatened, increased access to and use of reactive nitrogen can itself be an effective adaptation strategy, giving a fertiliser boost to crop growth rates and overall production that counterbalances the increasing losses from impacts such as damage from pests and disease[10]. In such cases, however, the nitrogen use efficiency may be low and, as climate change impacts intensify, the degree to which more fertiliser can offset them will rapidly shrink.

Interventions that directly increase resilience to climate change impacts may be much more sustainable and result in a large increase in nitrogen use efficiency, rather than a decrease. Examples include the planting of crop types that have greater resistance to drought or that flower at times of the day that avoid peak temperatures, thereby increasing pollination success rates[11]. Likewise, pest damage can be reduced by the use of companion crop planting, such as the 'push-pull' method used to control stem borer attacks in cereal crop production[12]. In this method the cereal crop is interspersed with companion plants like *Desmodium* that emit volatile compounds which deter pests (the 'push'), while the borders of the field are planted with grasses that attract the pests away from the main crop (the 'pull'). The *Desmodium* companion planting also helps to control weeds and, as it is a nitrogen-fixer, can also provide a useful supply of reactive nitrogen to the soil. Agroforestry – growing crops alongside trees – is another important climate change adaptation strategy in many tropical and sub-tropical regions[13]. In this system the trees can provide shade for the crops to help protect them from heat stress, with 'shade coffee' being a widely used example of this strategy[14]. Nitrogen-fixing trees and shrubs are often used to deliver the additional benefit of reactive nitrogen inputs to the soil[15]. Ultimately, the increased resilience to climate change impacts that such adaptation strategies provide allows for less fertiliser use per tonne of food produced – a 'win-win' of resilience and reduced nitrogen pollution.

In the developed world, and particularly in temperate and high-latitude regions, more intense rainfall and more frequent flooding can pose a major threat to crop yields[16,17]. In these systems threats from pests, disease and weeds tend to be controlled by pesticides and herbicides, while drought risks can often be ameliorated by the use of irrigation. Too much water, however, can be a much more difficult challenge and can have especially damaging impacts via its interactions with nitrogen.

Intense rainfall events can result in large losses of nitrogen from the soil via surface run-off and leaching[18]. These losses can be very large when heavy rainfall occurs soon after the application of nitrogen fertiliser or manure and where the soil is already very wet. With climate change projected to increase the intensity of rainfall events and, in many temperate areas, to also increase the total amount of rainfall in winter, losses of agricultural nitrogen to surface and groundwaters risk[19] becoming a major source of economic and environmental damage[20].

Many of the most effective strategies to adapt to this increasing risk of run-off and leaching involve improving the nitrogen use efficiency of crops and the alteration of farm management practices (see Chapter 9). Precision fertiliser application, crop nitrogen budgets, manure injection and the avoidance of nitrogen addition to waterlogged soils can all be highly effective ways to cut losses in the face of more extreme precipitation patterns[21].

Despite such efforts to increase farm nitrogen use efficiency and reduce how much is lost through leaching and run-off, increasingly intense rainfall events will inevitably lead to some losses to freshwater systems. Steps to decrease the level of nitrogen pollution from run-off and leaching have already been taken in many countries. In Europe, the European Union's wide-ranging Water Framework Directive aims to cut nitrogen pollution in ground and surface waters, while the Nitrates Directive sets specific targets for reducing water pollution by nitrate from agricultural land[22]. Large areas of the UK have now been designated nitrate vulnerable zones (NVZs) as a result of the high levels of nitrate in surface and ground waters[23]. Farms in these areas are required to follow a 'code of good practice', detailing the amount of nitrogen-based fertiliser that can be applied and when applications should take place. Unfortunately, achieving such pollution targets has proved problematic in the UK and elsewhere due to wide variations in soil type, increased intensification and scientific uncertainties[24]. With Northern Europe set to experience an increase in annual rainfall, and particularly in winter precipitation[25], due to climate change, these targets are likely to become even more difficult to meet. One key set of solutions to the nitrogen leaching and run-off problem focuses not on stopping the initial losses from farms, but instead on intercepting the nitrogen before it can pass too far downstream. Two of the most widely used strategies of this type in Europe are 'buffer strips' and 'constructed wetlands' (Box 13.1).

Box 13.1

Intercepting farm nitrogen loss

Buffer strips

For interception of nitrogen lost via leaching and surface run-off, vegetated buffer strips have the potential to greatly reduce the water pollution impacts of more intense rainfall events. Buffer strips are most commonly created alongside farm streams and drainage channels – called 'riparian' buffer strips – so that they can intercept nutrients, pathogens and sediments before they escape downstream from the farm[26]. Some are simply 5–10 metres wide grassed strips that are left uncultivated, while others are deliberately planted with shrubs and trees. They work both as physical and biological filters. First, the plants, soil and roots slow down water flow and trap sediments that have been washed from the farm soil. Next, the plant roots and soil microbes use the inflowing reactive nitrogen for growth and, in the case of the soil microbes, also as a source of energy. Where the buffer strip is wide enough and the vegetation well-established, more than half of the reactive nitrogen that would have entered the drainage streams can be intercepted. If trees are also planted as part of the buffer strip, then they can increase both nitrogen interception and carbon sequestration even further[27]. As the main mechanism by which soil microbes in buffer strips intercept reactive nitrogen is denitrification, and a by-product of denitrification is nitrous oxide, there may be some climate penalty to pay in terms of net climate forcing[28]. However, as a means to address nitrogen pollution of the aquatic environment and at the same time increase on-farm carbon sequestration buffer strips can represent a real 'win-win' in addressing the nitrogen and climate change challenge.

(Continued)

Box 13.1 (Continued)

Constructed wetlands

Constructed wetlands can be used on farms to intercept nitrogen after it has escaped into drainage waters, but before it can escape further downstream. During very intense rainfall events, constructed wetlands can also increase a farm's physical storage capacity for surface run-off – accommodating the surge in nutrient-rich water flowing from hard standings and livestock sheds. The nitrogen is intercepted by a combination of nitrification, plant uptake of ammonium and denitrification in the wetland soils[29]. The use of denitrifying wetlands and buffers has become increasingly widespread in the last few years, but some doubts remain over their suitability for all areas[30]. As well as the potential problem of leached nitrogen bypassing the wetland soils at times of high flow, there is again a risk that denitrification in their soils – often the primary process of nitrate removal – will lead to increased nitrous oxide emissions, thereby swapping a water pollution problem for a climate change issue[31] (see Chapter 9).

Camellones and climate change

In other regions of the world, similar problems of increasing nitrogen loss via surface run-off due to climate change pose a very serious risk to crop yields. For instance, where nutrient availability is already low, a more intense wet season may result in so much nitrogen being flushed from the soil that it is then unable to sustain cultivation. In Bolivia a highly successful adaptation response to this problem has been developed from a farming technique called 'camellones'[32]. Camellones are believed to date back to pre-Colombian civilisation circa 1,000 BC, when they were used to meet the challenge of growing food in the face of El Niño-induced swings in the length of wet and dry seasons. They involve the construction of closed-end canals and raised banks, whereby water is retained within the canals during the dry season to aid with irrigation and then, during the wet season, the raised banks provide non-waterlogged soil for cultivation[33]. Importantly, these canals and raised banks mean that soil nutrients are retained instead of being washed away during intense rainfall. With climate change projected to increase the intensity and variability of droughts and floods in this region, the camellones represent an effective adaptation strategy that ensures soil nitrogen levels and overall fertility are maintained.

Livestock production

Climate change is projected to have negative impacts on livestock agriculture in many regions[34], with implications for nitrogen use and losses. As mentioned earlier, more intense rainfall events may cause an increase in run-off from livestock hard standings and housing[35] – with potentially serious consequences with regard to water quality and eutrophication. As well as the buffer strip and constructed wetlands approaches (Box 13.1) to deal with increased livestock nitrogen run-off, installation of covered manure storage can be an effective way to avoid large losses to aquatic systems from intensive livestock production[36].

The spread of some pests and diseases may also serve to reduce nitrogen use efficiency in livestock farming[37], with reduced productivity or increased mortality meaning more nitrogen is required to produce each kilogram of meat or litre of milk. In the middle and lower latitudes, heat stress may become an increasing threat to livestock agriculture and its efficiency[38]. Provision of shading by trees has been a successful response in some areas[39], with the livestock benefiting from lower temperatures and the farmer able to gain additional income from timber or tree fruit production. For indoor livestock, the provision of sprinklers and good ventilation during the hottest months can yield benefits in terms of animal welfare and productivity[40].

Another consequence of the higher temperatures associated with climate change is an increase in the rate of ammonia volatilisation from livestock manure and urine. Injection of fertiliser and manure nitrogen can reduce these losses[41], but in many livestock systems, such as grazed or housed animal production, manure and urine inevitably end up being deposited on the surface of the ground. Here, a potential 'win-win' of climate change adaptation and mitigation exists in the form of on-farm tree planting and so-called 'canopy capture' (Box 13.2).

Supply-chain climate change adaptation and nitrogen

An important, though oft-overlooked, impact of climate change on agricultural nitrogen efficiency is through its effects on the food supply chain. Higher temperatures will themselves increase food spoilage rates and consequently risk increasing food loss and waste between farm and consumer[45]. Any impact that increases such losses will then require additional food production and incur the additional nitrogen use and losses associated with it. Increased provision of refrigeration, improved processing and packaging, and accelerated transfer of food from farms to retailers can all help limit the impact of additional warming on food spoilage[46].

Box 13.2

Canopy capture and ammonia

In intensive livestock systems, such as cattle feedlots and barn egg production, reactive nitrogen emissions to the atmosphere in the form of ammonia can be very substantial[42]. Once emitted this ammonia may travel long distances and cause air pollution issues while it remains in the atmosphere, or eutrophication and acidification problems where it is redeposited to land or water (see Chapter 5). By careful placing of trees directly above or downwind of the ammonia source it is possible to create a physical buffer for these emissions, with the tree canopy trapping the ammonia[43]. This trapped reactive nitrogen can then also help to enhance carbon growth and its uptake by the 'buffer' trees. Depending on the type of farming system, creating a band of trees downwind of the ammonia source instead of having a canopy directly above the livestock may be most appropriate. Where livestock heat stress is an issue, the provision of shade trees in the grazed areas can provide the triple role of cooling livestock, reducing ammonia volatilisation rates and intercepting a proportion of any ammonia that is lost.

Whether planted as downwind 'bands' or as a direct canopy above livestock, if correctly placed and with the right characteristics in terms of canopy height and density, these tree buffers can intercept a large proportion of ammonia lost to the air from intensive poultry systems[43]. With climatic warming combined with an increase in global livestock production likely to mean spiraling ammonia emissions from agriculture, such integration of tree planting has real potential to limit emissions. Assuming such planting of 'capture' trees also provided a net increase in carbon stocks, a double climate change mitigation benefit could be delivered through reduced nitrous oxide emissions from redeposited ammonia and increased carbon dioxide uptake[44].

Severe weather impacts induced by climate change, such as storm and flood damage to food storage and transport links, also represent a significant risk in terms of pushing up food loss and wastage rates around the world[47,48]. The provision of more food storage facilities that are resilient to extreme weather conditions, such as high-spec grain silos, may be an effective solution, but bring with them additional costs to farmers. Likewise, the use of alternative supply routes in the event of primary river, road or rail links being cut is limited in many areas by a dearth of options and a shortage of finance for new transport infrastructure.

As climate change will have different impacts depending on place and time, its impacts on nitrogen fluxes and process will also vary with time and space. With food and resource supplies being increasingly connected around the world, any severe impact in one area can have repercussions hundreds or even thousands of miles away[49]. In the agricultural sector these linkages are especially important in the fight to improve nitrogen use efficiency. Where climate change impacts reduce food production to a great enough extent, they can push up demand globally[50,51] leading to the cultivation of this food in less optimal areas where nitrogen use efficiency is much lower and the losses much higher. Severe weather impacts, such as heatwave-induced losses in wheat production in Russia and the US, have already shown how global commodity prices (and consequently production incentives) can be affected in this way[52].

Integrated adaptation

Invariably, where resilience to climate change can be increased, so too can nitrogen use efficiency and the myriad benefits that go with it. In facing the global challenge that is climate change, nitrogen can become either a great ally or a formidable foe. Ensuring it is the former will require the integrated understanding and evidence-based policy making that has so often been lacking to date.

Airborne nitrogen

As well as the many interactions of climate change impacts with agricultural nitrogen, projected changes in temperature, precipitation and severe weather events are likely to alter atmospheric nitrogen fluxes and processes much more widely. In the atmosphere, high summer temperatures and more intense high-pressure weather systems may result in dangerous increases in low-level ozone concentrations – with nitrogen oxide (NOx) gases being a key precursor[53]. In many towns and cities, the urban heat island (UHI) effect combined with climate warming and NOx emissions from transport may induce serious human health risks. Reducing emissions from transport is an effective direct response to this risk, but reducing urban temperatures – though the use of reflective roofs and surfaces or by increasing urban vegetation – can also help to offset the impacts of a warming climate[54].

Outside of urban centres, reducing NOx emissions from fossil fuel combustion by power plants may be made more difficult due to the impacts of climate change. Where electricity is generated by 'thermal' plants – coal and gas powered steam turbines – higher temperatures will mean a drop in efficiency[55]. To bridge this energy gap more coal or gas will then have to be burned, leading to yet more NOx emission unless additional flue gas scrubbing is put in place.

The increased frequency and intensity of wildfires that is projected as an impact of climate change in the 21st century may also result in a rise in nitrogen losses to the atmosphere, mainly in the form of NOx and nitrous oxide[56]. Improved fire management and suppression may play an important role in limiting these emissions, with the control of peatland fires in Asia being of particular importance to global nitrogen emissions from wildfires[57].

Freshwater nitrogen

For freshwater systems around the world, the increasing threat of drought in many regions is likely to result in a deliberate expansion of water storage capacity, such as in reservoirs, and in more wastewater reclamation. This adaptation response may drastically alter the freshwater nitrogen processing of catchments, with potentially significant implications for nitrous oxide emissions[58].

Greater nitrogen and phosphorous losses to freshwater systems caused by intense rainfall events may themselves exacerbate flooding problems, with the lost nutrients going on to boost plant growth in streams, rivers and drainage channels, thereby causing more frequent blockages and floods[59]. Such increased eutrophication of freshwater catchments will therefore require increased clearance efforts, on top of those required to deal with the increasingly frequent and extreme surges in stream water flows.

Marine nitrogen

The increases in storm water run-off and river flows projected for the 21st century in temperate and high-latitude regions will put ever-greater pressure on sewage processing systems around the world. As processing capacity is exceeded, nitrogen-rich effluent may be more frequently released into estuarine and coastal waters[60]. The eutrophication and harmful algal bloom (HAB) problems that result from this may incur substantial economic, environmental and human health costs (see Chapter 8). Development of sewage storage and processing facilities with the capacity to cope with the projected increases in precipitation will be crucial in limiting these negative impacts.

References

1. Stocker, T. *Climate change 2013: the physical science basis: Working Group I contribution to the Fifth assessment report of the Intergovernmental Panel on Climate Change.* (Cambridge University Press, 2014).
2. Parry, M. L., Rosenzweig, C., Iglesias, A., Livermore, M. & Fischer, G. Effects of climate change on global food production under SRES emissions and socio-economic scenarios. *Global Environmental Change* **14**, 53–67 (2004).
3. Piao, S. et al. The impacts of climate change on water resources and agriculture in China. *Nature* **467**, 43–51 (2010).
4. Howden, S. M. et al. Adapting agriculture to climate change. *Proceedings of the National Academy of Sciences of the United States of America* **104**, 19691–19696 (2007).
5. Thornton, P. K., Jones, P. G., Alagarswamy, G. & Andresen, J. Spatial variation of crop yield response to climate change in East Africa. *Global Environmental Change* **19**, 54–65 (2009).
6. Galloway, J. N. et al. The nitrogen cascade. *Bioscience* **53**, 341–356 (2003).
7. Erisman, J. W., Sutton, M. A., Galloway, J., Klimont, Z. & Winiwarter, W. How a century of ammonia synthesis changed the world. *Nature Geoscience* **1**, 636–639, doi:10.1038/ngeo325 (2008).
8. Sutton, M. A. et al. *The European nitrogen assessment: sources, effects and policy perspectives.* (Cambridge University Press, 2011).
9. Lobell, D. B. et al. Prioritizing climate change adaptation needs for food security in 2030. *Science* **319**, 607–610 (2008).
10. Denning, G. et al. Input subsidies to improve smallholder maize productivity in Malawi: toward an African Green Revolution. *PLoS Biology* **7**, e1000023 (2009).
11. Wassmann, R. et al. Climate change affecting rice production: the physiological and agronomic basis for possible adaptation strategies. *Advances in Agronomy* **101**, 59–122 (2009).
12. Hassanali, A., Herren, H., Khan, Z. R., Pickett, J. A. & Woodcock, C. M. Integrated pest management: the push–pull approach for controlling insect pests and weeds of cereals, and its potential for other agricultural systems including animal husbandry. *Philosophical Transactions of the Royal Society B: Biological Sciences* **363**, 611–621 (2008).
13. Verchot, L. V. et al. Climate change: linking adaptation and mitigation through agroforestry. *Mitigation and Adaptation Strategies for Global Change* **12**, 901–918 (2007).
14. Lin, B. B. Agroforestry management as an adaptive strategy against potential microclimate extremes in coffee agriculture. *Agricultural and Forest Meteorology* **144**, 85–94 (2007).
15. Garrity, D. P. et al. Evergreen Agriculture: a robust approach to sustainable food security in Africa. *Food Security* **2**, 197–214 (2010).
16. Olesen, J. E. & Bindi, M. Consequences of climate change for European agricultural productivity, land use and policy. *European Journal of Agronomy* **16**, 239–262 (2002).
17. Falloon, P. & Betts, R. Climate impacts on European agriculture and water management in the context of adaptation and mitigation – the importance of an integrated approach. *Science of the Total Environment* **408**, 5667–5687 (2010).

18. Hessen, D. O., Hindar, A. & Holtan, G. The significance of nitrogen runoff for eutrophication of freshwater and marine recipients. *Ambio* **26**, 312–320 (1997).
19. Andersen, H. E. et al. Climate-change impacts on hydrology and nutrients in a Danish lowland river basin. *Science of the Total Environment* **365**, 223–237 (2006).
20. Grizzetti, B. et al. Nitrogen as a threat to European water quality. In *The European Nitrogen Assessment*, edited by M. Sutton et al., 379–404 (Cambridge University Press, UK, 2011).
21. Raun, W. R. & Johnson, G. V. Improving nitrogen use efficiency for cereal production. *Agronomy Journal* **91**, 357–363 (1999).
22. Goodchild, R. EU policies for the reduction of nitrogen in water: the example of the Nitrates Directive. *Environmental Pollution* **102**, 737–740 (1998).
23. Osborn, S. & Cook, H. F. Nitrate vulnerable zones and nitrate sensitive areas: a policy and technical analysis of groundwater source protection in England and Wales. *Journal of Environmental Planning and Management* **40**, 217–234 (1997).
24. Tilman, D., Cassman, K. G., Matson, P. A., Naylor, R. & Polasky, S. Agricultural sustainability and intensive production practices. *Nature* **418**, 671–677 (2002).
25. Rounsevell, M. D. A. & Reay, D. S. Land use and climate change in the UK. *Land Use Policy* **26**, S160–S169, doi:10.1016/j.landusepol.2009.09.007 (2009).
26. Hefting, M. M. & de Klein, J. J. Nitrogen removal in buffer strips along a lowland stream in the Netherlands: a pilot study. *Environmental Pollution* **102**, 521–526 (1998).
27. Fortier, J., Gagnon, D., Truax, B. & Lambert, F. Nutrient accumulation and carbon sequestration in 6-year-old hybrid poplars in multiclonal agricultural riparian buffer strips. *Agriculture, Ecosystems & Environment* **137**, 276–287 (2010).
28. Hefting, M. M., Bobbink, R. & de Caluwe, H. Nitrous oxide emission and denitrification in chronically nitrate-loaded riparian buffer zones. *Journal of Environmental Quality* **32**, 1194–1203 (2003).
29. Mitsch, W. J. et al. Reducing Nitrogen Loading to the Gulf of Mexico from the Mississippi River Basin: Strategies to Counter a Persistent Ecological Problem Ecotechnology – the use of natural ecosystems to solve environmental problems – should be a part of efforts to shrink the zone of hypoxia in the Gulf of Mexico. *BioScience* **51**, 373–388 (2001).
30. Dosskey, M. G. Setting priorities for research on pollution reduction functions of agricultural buffers. *Environmental Management* **30**, 641–650 (2002).
31. Reay, D. S. Fertilizer 'solution' could turn local problem global – protecting soil and water from pollution may mean releasing more greenhouse gas. *Nature* **427**, 485 doi:10.1038/427485a (2004).
32. Parry, M. L. *Climate Change 2007: impacts, adaptation and vulnerability: contribution of Working Group II to the fourth assessment report of the Intergovernmental Panel on Climate Change*. Vol. 4. (Cambridge University Press, 2007).
33. Biesboer, D. D., Binford, M. W. & Kolata, A. Nitrogen fixation in soils and canals of rehabilitated raised-fields of the bolivian altiplano. *Biotropica* **31**, 255–267 (1999).

34. Thornton, P., Van de Steeg, J., Notenbaert, A. & Herrero, M. The impacts of climate change on livestock and livestock systems in developing countries: a review of what we know and what we need to know. *Agricultural Systems* **101**, 113–127 (2009).
35. Edwards, A. C. et al. Farmyards, an overlooked source for highly contaminated runoff. *Journal of Environmental Management* **87**, 551–559 (2008).
36. Oenema, O., Oudendag, D. & Velthof, G. L. Nutrient losses from manure management in the European Union. *Livestock Science* **112**, 261–272 (2007).
37. Purse, B. V. et al. Climate change and the recent emergence of bluetongue in Europe. *Nature Reviews Microbiology* **3**, 171–181 (2005).
38. West, J. Effects of heat-stress on production in dairy cattle. *Journal of Dairy Science* **86**, 2131–2144 (2003).
39. Payne, W. A review of the possibilities for integrating cattle and tree crop production systems in the tropics. *Forest Ecology and Management* **12**, 1–36 (1985).
40. Mader, T. L. Environmental stress in confined beef cattle. *Journal of Animal Science* **81**, E110–E119 (2003).
41. Sommer, S. G. & Hutchings, N. Ammonia emission from field applied manure and its reduction – invited paper. *European Journal of Agronomy* **15**, 1–15 (2001).
42. Meisinger, J. & Jokela, W. Ammonia volatilization from dairy and poultry manure. Managing nutrients and pathogens from animal agriculture. *Natural Resource, Agriculture, and Engineering Service, Ithaca, NY*, **NRAES-130**, 334–354 (2000).
43. Patterson, P. et al. The potential for plants to trap emissions from farms with laying hens. 1. Ammonia. *The Journal of Applied Poultry Research* **17**, 54–63 (2008).
44. Malone, G. & Van Wicklen, G. Trees as a vegetative filter. *Poultry Digest Online* **3**, 7 (2001).
45. Parfitt, J., Barthel, M. & Macnaughton, S. Food waste within food supply chains: quantification and potential for change to 2050. *Philosophical Transactions of the Royal Society B: Biological Sciences* **365**, 3065–3081 (2010).
46. James, S. & James, C. The food cold-chain and climate change. *Food Research International* **43**, 1944–1956 (2010).
47. Vermeulen, S. J., Campbell, B. M. & Ingram, J. S. Climate change and food systems. *Annual Review of Environment and Resources* **37**, 195 (2012).
48. Haile, M. Weather patterns, food security and humanitarian response in sub-Saharan Africa. *Philosophical Transactions of the Royal Society B: Biological Sciences* **360**, 2169–2182 (2005).
49. Hanjra, M. A. & Qureshi, M. E. Global water crisis and future food security in an era of climate change. *Food Policy* **35**, 365–377 (2010).
50. O'Brien, K. L. & Leichenko, R. M. Double exposure: assessing the impacts of climate change within the context of economic globalization. *Global Environmental Change* **10**, 221–232 (2000).
51. Battisti, D. S. & Naylor, R. L. Historical warnings of future food insecurity with unprecedented seasonal heat. *Science* **323**, 240–244 (2009).
52. Teixeira, E. I., Fischer, G., van Velthuizen, H., Walter, C. & Ewert, F. Global hot-spots of heat stress on agricultural crops due to climate change. *Agricultural and Forest Meteorology* **170**, 206–215 (2013).

53. Meleux, F., Solmon, F. & Giorgi, F. Increase in summer European ozone amounts due to climate change. *Atmospheric Environment* **41**, 7577–7587 (2007).
54. Rosenfeld, A. H., Akbari, H., Romm, J. J. & Pomerantz, M. Cool communities: strategies for heat island mitigation and smog reduction. *Energy and Buildings* **28**, 51–62 (1998).
55. van Vliet, M. T. et al. Vulnerability of US and European electricity supply to climate change. *Nature Climate Change* **2**, 676–681 (2012).
56. Jaffe, D. A. & Wigder, N. L. Ozone production from wildfires: a critical review. *Atmospheric Environment* **51**, 1–10 (2012).
57. Levine, J. S. *Biomass burning and its inter-relationships with the climate system* 15–31 (Springer, 2000).
58. Baron, J. et al. The interactive effects of excess reactive nitrogen and climate change on aquatic ecosystems and water resources of the United States. *Biogeochemistry* **114**, 71–92 (2013).
59. O'Hare, M. T. et al. Eutrophication impacts on a river macrophyte. *Aquatic Botany* **92**, 173–178 (2010).
60. Semadeni-Davies, A., Hernebring, C., Svensson, G. & Gustafsson, L.-G. The impacts of climate change and urbanisation on drainage in Helsingborg, Sweden: combined sewer system. *Journal of Hydrology* **350**, 100–113 (2008).

Conclusion

The preceding chapters have highlighted how past and current management of nitrogen in a changing climate has often failed. To date, humankind's attempts at effectively managing nitrogen have been exemplified by reactive and patchy responses that focus too much on one form of nitrogen in one place. When it comes to effective management of nitrogen, its myriad forms and the global distribution of their impacts and interactions with climate change make for a Herculean task.

On the local scale, farmers might try hard to give precisely the right nitrogen supply to their crops, only for more intense rainstorms to wash away much of the added fertiliser. For individual nations, tight regulations might be put in place to cut nitrogen pollution (such as NOx gases) from power stations, only for the prevailing winds to bring in a wave of nitrogen-polluted air from a neighbouring country. Even at the global level the issues of nitrogen and climate change make for a policy maker's nightmare. Nitrogen can slip so easily between different forms and can also be transported huge distances. With some forms of nitrogen posing a water pollution issue, others degrading air quality and some being powerful drivers of the greenhouse effect, a policy aimed at addressing one nitrogen issue may end up exacerbating another (e.g. injecting fertilisers to reduce volatilisation to the air can increase leaching losses to water).

Global nitrogen management

Co-ordination of policies within and between the regions of the world is what is so badly lacking when it comes to nitrogen. Though there are frameworks in place covering some nitrogen issues for some areas – such as the Helsinki Commission to curb eutrophication of the Baltic Sea or the UN's Convention on Biological Diversity (CBD) to consider impacts on biodiversity – a holistic framework for global nitrogen management is still lacking.

A global policy framework for nitrogen management is now needed that can co-ordinate the response to the multiple threats and opportunities of nitrogen in a fully joined-up way. This framework would deliver nitrogen policy recommendations that were locally and regionally applicable, while still being consistent with policies in other nations and regions of the world. Such a framework would allow transboundary issues, such as NOx pollution from fossil fuel burning, to be tackled much more effectively. It would also help to expose the antagonisms and issues related to instances when a mitigation option intended to reduce one pollutant causes an increase in a different pollutant (called 'pollution swapping').

As nitrogen is also a key player in the challenges related to ozone depletion, food supply, energy security and in tackling climate change itself, any such attempt to effectively manage nitrogen at a global level will also need access to the very best evidence from across a host of stakeholders, sectors and locations. Following the model of the Intergovernmental Panel on Climate Change (IPCC) in terms of the provision of robust, policy-neutral science, a co-ordination of nitrogen research, measurement and modelling work that encompasses air, soil and water systems around the world will be required.

The beginnings of such a research network are already forming – the International Nitrogen Initiative (INI) being an exemplar – and, if the usual academic silos of disciplinarity and geographic location can be overcome, real progress can be made on getting the best nitrogen science into the best policy.

Our knowledge of nitrogen and climate change may be far from complete – the complexities are certainly great – but the threats and opportunities these twin issues hold for humankind in the 21st century demand that future policy achieves much more than chasing nitrogen shadows.

Recommended Reading

This book is focused on the interactions between nitrogen and climate change – an area of scientific understanding that has seen great strides forward in recent decades but where big uncertainties still remain. The below represent, in my view, some of the best and most comprehensive books and papers to date dealing with nitrogen's role in global change and human civilization.

Erisman, J. W., M. A. Sutton, J. Galloway, Z. Klimont and W. Winiwarter (2008). "How a century of ammonia synthesis changed the world." *Nature Geoscience* 1(10): 636–639.

Galloway, J. N., A. R. Townsend, J. W. Erisman, M. Bekunda, Z. Cai, J. R. Freney, L. A. Martinelli, S. P. Seitzinger and M. A. Sutton (2008). "Transformation of the nitrogen cycle: recent trends, questions, and potential solutions." *Science* 320(5878): 889–892.

Smil, V. (2001). *Enriching the earth : Fritz Haber, Carl Bosch, and the transformation of world food production.* Cambridge, MA: MIT Press.

Smith, K. A. (2010). *Nitrous oxide and climate change.* London: Earthscan.

Suddick, E. and E. Davidson (2012). "The role of nitrogen in climate change and the impacts of nitrogen-climate interactions on terrestrial and aquatic ecosystems, agriculture, and human health in the United States: a technical report submitted to the US National Climate Assessment." *North American Nitrogen Center of the International Nitrogen Initiative (NANC-INI), Woods Hole Research Center* 149: 208.

Sutton, M. A., C. M. Howard, J. W. Erisman, G. Billen, A. Bleeker, P. Grennfelt, H. van Grinsven and B. Grizzetti (2011). *The European nitrogen assessment: sources, effects and policy perspectives.* Cambridge: Cambridge University Press.

Index

Page numbers in **bold** refer to figures and *italics* refer to tables.

Lightning Source UK Ltd.
Milton Keynes UK
UKOW05f2114020615

252785UK00006B/14/P